Establishing Site-Specific Flushing Velocities

Prepared by:
Melinda J. Friedman, **Kathy Martel**, and **Andrew Hill**
Economic and Engineering Services, Inc.
and
David Holt, **Sian Smith**, **Twan Ta**, and **Chris Sherwin**
Thames Water Utilities, Ltd.
and
David Hiltebrand, **Peter Pommerenk**, and **Zeina Hinedi**
AH Environmental Consultants, Inc.
and
Anne Camper
Montana State University

Jointly sponsored by:
Awwa Research Foundation
6666 West Quincy Avenue, Denver, CO 80235-3098
and
Kiwa N.V.
Research and Consultancy
Goningenhaven 7, P. O. Box 1072, 3430 B B Nieuwegein, The Netherlands

Published by:

Awwa Research Foundation
Advancing the Science of Water®

IWA Publishing

DISCLAIMER

Published by IWA Publishing, Alliance House, 12 Caxton Street, London SW1H 0QS, UK

Telephone: +44 (0) 20 7654 5500; Fax: +44 (0) 20 7654 5555; Email: publications@iwap.co.uk;
Web: **www.iwapublishing.com**

AwwaRF report number 90964F. Originally published by AwwaRF for its subscribers in 2003.
IWA Publishing version published 2004.

ISBN 1 84339 870 2

CONTENTS

TABLES

FIGURES

FOREWORD

The Awwa Research Foundation is a nonprofit corporation that is dedicated to the implementation of a research effort to help utilities respond to regulatory requirements and traditional high-priority concerns of the industry. The research agenda is developed through a process of consultation with subscribers and drinking water professionals. Under the umbrella of a Strategic Research Plan, the Research Advisory Council prioritizes the suggested projects based upon current and future needs, applicability, and past work; the recommendations are forwarded to the Board of Trustees for final selection. The foundation also sponsors research projects through the unsolicited proposal process; the Collaborative Research, Research Applications, and Tailored Collaboration programs; and various joint research efforts with organizations such as the U.S. Environmental Protection Agency, the U.S. Bureau of Reclamation, and the Association of California Water Agencies.

This publication is a result of one of these sponsored studies, and it is hoped that its findings will be applied in communities throughout the world. The following report serves not only as a means of communicating the results of the water industry's centralized research program but also as a tool to enlist the further support of the nonmember utilities and individuals.

Projects are managed closely from their inception to the final report by the foundation's staff and large cadre of volunteers who willingly contribute their time and expertise. The foundation serves a planning and management function and awards contracts to other institutions such as water utilities, universities, and engineering firms. The funding for this research effort comes primarily from the Subscription Program, through which water utilities subscribe to the research program and make an annual payment proportionate to the volume of water they deliver and consultants and manufacturers subscribe based on their annual billings. The program offers a cost-effective and fair method for funding research in the public interest.

A broad spectrum of water supply issues is addressed by the foundation's research agenda: resources, treatment and operations, distribution and storage, water quality and analysis, toxicology, economics, and management. The ultimate purpose of the coordinated effort is to assist water suppliers in providing the highest possible quality of water economically and reliably. The true benefits are realized when the results are implemented at the utility level. The foundation's trustees are pleased to offer this publication as a contribution toward that end.

Edmund G. Archuleta, P.E.
Chair, Board of Trustees
Awwa Research Foundation

James F. Manwaring, P.E.
Executive Director
Awwa Research Foundation

ACKNOWLEDGMENTS

The authors would like to acknowledge the gracious financial support of the Awwa Research Foundation and KIWA. Without their support, this project would not have been possible.

This research has been guided from the outset by an experienced Project Advisory Committee (PAC) who provided valuable guidance, expertise and suggestions. The authors wish to thank these individuals for their time and efforts:

Faye Cossins, Cincinnati Water Works, Cincinnati, OH
Jeff Farrell, BHC Company, Bridgeport, CT
Kan Oberoi, Charleston Commissioners of Public Works, Charleston, SC
Harry Romgens, Kiwa, The Netherlands

A number of water utilities offered their time, staff expertise, and facilities for case studies and field investigations. The authors gratefully acknowledge these efforts:

Anchorage Water and Wastewater Utility, AK
City of Portland, OR
City of Cincinatti, OH
City of Cleveland, OH
City of Newport News, VA
Irvine Ranch Water District, CA
City of Bellevue, WA
South Berwick Water District, ME
City of Dover, NH
Greater Vancouver Regional District, British Columbia, Canada
Southern California Water Company, CA
Thames Water Company, UK
Yarra Valley Water, Australia

The authors thank the following staff of Economic and Engineering Services, Inc. for their efforts in bringing this work together into its final form:

Shari Neuenschwander, Office Manager
Druann O'Connor, Administrative Assistant

The findings of the CFD modeling show that the most significant factors in determining the flush rate are the extent of pipe tuberculation and particle density. Particle size is less significant and the pipe diameter has almost no effect for the range of pipe diameters analyzed (4-inch to 8-inch). Specific findings from CFD modeling include:

(1) Pipe tuberculation has a significant effect upon flushing, reducing the achievable removal performance. Increasing pipe tuberculation makes flushing performance less sensitive to flushing velocity.
(2) Particle specific gravity has a significant effect upon removal. More dense particles require higher average mean velocities for lifting.
(3) Nominal pipe diameter does not significantly affect particle removal performance for 4-inch (100-mm), 6-inch (150-mm), or 8-inch (200-mm) mains.
(4) The removal of larger particles requires slightly higher flushing velocities. For example, 200-μm particles required up to 0.66 fps (0.2 m/s) higher mean velocity compared to 20-μm particles.
(5) Current industry flush rates for loose deposits are higher than required.

The CFD results predict significantly better particle removal than those of Stephenson (1989). For particles of specific gravity of 1.5 and 200 μm size in a 4-inch (100-mm) main, Stephenson (1989) recommends a minimum flushing velocity of 3.0 fps (0.9 m/s) for particle removal. In contrast, CFD modeling illustrated that flow velocities below 1.3 fps (0.4 m/s) would be sufficient for complete removal of these particles from smooth mains. For particles of specific gravity of 3.0 and 200 μm size in a 4-inch (100-mm) main, Stephenson (1989) recommends a minimum flushing velocity of 5.9 fps (1.8 m/s) for particle removal. However, CFD modeling illustrated that flow velocities of 2.0 fps (0.6 m/s) were sufficient to scour these particles from smooth surfaces.

The models developed around the smooth pipe agree well with the 'rule-of-thumb' value provided by Williams (2000). This rule-of-thumb, which is based upon previous pipe rig data, suggests that a mean velocity of 1.6 fps (0.5 m/s) is required to suspend most loose sediment. However, the CFD data suggests it should not be used for calculating flush rates for tuberculated pipes.

Pilot Study Program

The aim of the pilot study program was to determine, in a controlled environment, the flushing velocities required for lifting and removal of different types of contaminants. Materials utilized in the pilot studies include loose deposits, cohesive deposits, and adherent deposits. A second objective was to compare the lifting velocities derived from experimental testing to those derived from CFD modeling.

The pilot study trials have demonstrated that different types of contaminants have different degrees of adherence and cohesiveness, impacting the effectiveness of flushing as a control measure. Most of the contaminants tested would be removed by adopting flush rates determined by CFD modeling. Specific conclusions are cited below.

- **Sand, GAC, Iron corrosion particles.** These particles are commonly found in distribution systems and are amenable to removal by low level flushing at rates determined by CFD modeling.
- **Iron corrosion sludge.** This type of deposit can develop on unlined cast iron pipes. Iron sludge forms a very loose deposit and can be easily lifted, therefore amenable to removal by low level flushing. The flow rates for removal of loose deposits predicted by CFD modeling will remove iron sludge. The results indicate that iron sludge can accumulate in sufficient quantities to cause discoloration in parts of the network with unlined iron pipes where periodic stagnation occurs. In this scenario, there is a potential for discoloration and thus preventive flushing may be considered prior to any network activities involving hydraulic changes.
- **Manganese coating.** This type of deposit formed a very adherent coating on PVC pipe, not amenable to removal by flushing. Alternative, more aggressive control measures are required to remove this type of contaminant (e.g., swabbing, rehabilitation).
- **GAC fines**. Trials indicate that GAC can form a complex deposit in pipe networks. GAC deposits can exhibit loose or cohesive properties, influenced by factors such as material type and length of stagnation periods. The trials showed that for iron pipes, the GAC formed a loose deposit amenable to removal by flushing at flow rates predicted by CFD modeling. For PVC pipes, GAC formed a complex deposit on pipe walls and required a higher velocity to remove.
- **Ferric floc.** Ferric floc forms a very loose deposit and can be easily lifted, therefore amenable to removal by low level flushing. The flow rates for removal of loose deposits predicted by CFD modeling will remove ferric floc. Ferric floc is likely to cause discoloration events since it can be easily entrained and must be removed prior to any work involving hydraulic changes to the network.
- **Iron tubercles.** This type of deposit formed a very adherent coating in hard source waters not amenable to removal by flushing. Alternative, more aggressive, control measures may be required to remove this type of contaminant (e.g., swabbing, rehabilitation).

Table ES.1 summarizes the lifting velocities recorded in the pilot scale trials for the different types of contaminants.

Table ES.1
Summary of lifting velocities for different contaminants from experimental flushing trials

Contaminant Category	Contaminant	Range of lifting velocities recorded during different trials (fps)
Loose deposits	Sand	1.70 - 1.91
	GAC	1.48 - 1.71
	Iron corrosion	1.58 - 2.36
Cohesive deposits	Iron corrosion sludge	< 0.43
	Manganese coating	No lifting observed
	GAC clumps	0.98 - 3.94
	Ferric floc	0.69 - 0.92
Adherent deposits	Iron tubercles	No lifting observed

Field Trials

Distribution system flushing trials were conducted at 12 participating utilities. The purpose of the field trials was to examine the validity of a contaminant-specific flushing velocity approach in practice and to assess the effectiveness of flushing velocities in systems with varying pipe materials and source water qualities. Specific objectives included:

- Determine if more material was removed as velocity increased,
- Determine if different material was removed as velocity increased, and
- Identify what variables influenced or controlled the effectiveness of various flushing velocities.

The Project Team conducted unidirectional flushing for one to three loops at each utility. Each loop consisted of several steps requiring specific valve sequencing in order to direct water from a single clean source through a particular stretch of main. Most steps involved unidirectional flushing at a specific velocity and performing on-site analyses, including time-to-clear, turbidity, pH, and disinfectant residual measurements. These data were used to assess general water quality changes during the duration of a flushing event. In order to draw velocity-specific conclusions, sequential low and high velocity flushing trials were performed on select step(s) at each water system. The low and high velocity target ranges were 2-4.5 fps (0.6-1.4 m/s) and greater than 5 fps (1.5 m/s), respectively. Additional sampling conducted during these steps involved bulk water collection from the hydrant sample tap and particle capture from the discharge hose using hydrant nets.

Bulk water sample collection provided characterization and quantification of contaminants removed during unidirectional flushing. Samples were collected in laboratory-prepared containers and analyzed by various local commercial laboratories. The broader results of bulk water analyses were used to help answer the question: "Does a higher flushing velocity remove more material?"

The results of bulk water sampling and analysis provide some indication of the influence of different variables on deposit accumulation rates and flushing performance. However, since the magnitude of different variables involved is comparable to the size of the data set, a limited number of comparisons and conclusions can be drawn. The type of pipe material and presence or absence of an internal lining appears to play the most significant role with respect to the accumulation of ferric deposits removable via flushing. This conclusion is based on a comparison of normalized iron mass results from systems utilizing asbestos cement, ductile iron, cement-lined cast iron, and unlined cast iron. Those studies conducted on UCI mains consistently exhibited substantially higher accumulations of iron than comparable systems with different pipe materials and/or linings. Table ES.2 provides a summary of the efficiencies of the low velocity unidirectional flushing events for removing bulk water contaminants.

Table ES.2
Average low velocity removal efficiencies for bulk water trials

Pipe material	E_{Fe} (%)	E_{Mn} (%)
Asbestos cement	67.5%	65.0%
Polyvinyl chloride	66.5%	NT
Ductile iron	43.0%	39.0%
Unlined cast iron[1]	23.7%	46.7%

NT = Not tested
(1) Never been previously flushed

Particle capture trials were performed as part of the field tests in order to characterize distribution system deposits and evaluate flushing performance. The broader results of particle analyses were used to help answer the following two questions:

(1) "Does a higher flushing velocity remove more material?"
(2) "Does a higher flushing velocity remove different material?"

Hydrant nets were used to collect deposits at the point of flow discharge. The setup consisted of a 308-micron nylon net enclosed by a 57-micron nylon net. As such, conclusions cannot be drawn for particles smaller than 57 microns in size. The recovered particles were subject to laboratory testing using the following analyses:

- Fractionation (Sieving)
- Chemical Composition
- Bacteriological Composition
- Particle Density (Specific Gravity)

Bacteriological composition analysis consisted of measuring the density of HPC bacteria adhered to the recovered particles. However, the results were ultimately discarded. The Project Team concluded that if the bacteria were capable of being removed from the pipe surface, they were equally capable of being scoured from the particles as the water violently diffused through the nets. Further, the pore size of the fine net (57-micron) would permit the passage of several types of bacteria and silt particles capable of harboring bacteria.

Since the chemical formulae of distribution system deposits were not fully determined, particle size, elemental composition, and specific gravity were used to assess particle differences.

Table ES.3 provides qualitative responses to the above two questions. These results can be used by utilities to determine whether the benefits of high velocity flushing outweigh the costs.

Table ES.3
Qualitative evaluation of particle results

Site	Step	Total mass	Average particle size	Elemental composition	Specific gravity
ANC	7	SI	SI	NC	MI
BEL	2	MI	SI	NC	NA
CLE	12	SI	MI	NC	MD
NNI	8	SI	NC	SD	NA
POR	5	SI	NA	SI	NA

SI = Substantial increase from low to high velocity
MI = Moderate increase from low to high velocity
NC = Negligible change from low to high velocity
MD = Moderate decrease from low to high velocity
SD = Substantial decrease from low to high velocity
NA = Not Applicable

High velocity flushing removed substantially more material in 4 of the 5 field trials. However, based on size, composition, and density data, it appears as though the contaminants removed at low and high velocities were not always different. Perhaps the most important finding from the field trials is that "real-world" deposits can be loose, compacted, cohesive, or adherent, and these properties dictate the ability to scour and entrain specific particles.

STUDY LIMITATIONS AND DATA GAPS

The project's literature review showed that little published information was available on materials characterization. Most literature focused on adhered deposits due to their role in the passivation of the corrosion process. Research efforts to date have focused on describing the structure, chemical composition, and mechanisms of formation of pipe wall deposits. Relatively few studies address the adherence and size of pipe wall deposits. Of the few studies conducted on loose deposits in distribution systems, findings have focused on their characterization in terms of chemical composition and their occurrence in the distribution system. Little data are available on the physical characteristics of loose deposits, such as density and size. There is minimal published information addressing particles in the bulk water column. Examination of the nature of particles causing discolored water has been rare, perhaps due to difficulties analyzing what is often very small amounts of material or because the interpretation of the source of the particles is very difficult.

After development of several preliminary case studies as part of this Project, it became evident that the participating utilities had not selected flushing velocities based on site-specific conditions. The objective of the case studies was to capture information from utilities with long-term flushing programs and databases. Specific data of interest to this project included pre- and post-flushing water quality data, records of pipe ages/conditions, information on targeted particles, and records of flushing operations such as recorded pressures or flushing velocities.

The case studies sought to answer specific questions such as whether the utilities used targeted velocities, how they determined acceptable velocities, how they judged the success of flushing, and whether they had correlated pipe velocities to particular materials.

CFD modeling was conducted as part of this Project to predict loose particle behavior only. While this effort produced extremely useful and experimentally-verified data, the nature of deposits in many distribution systems is very complex, and may exist as loose, cohesive, compacted, and/or adhered particles. Additionally, CFD modeling results are presented over the velocity range of 1.3 to 4.6 fps. The database should be expanded to analyzed velocities up to 8 fps and for particle sizes up to 2 mm.

The Project's pilot trials allowed for much better control over flushing variables, and the use of pipe sections allowed for visual inspection of pipe surfaces before and after each flush. Pilot testing was used to explore the effects of pipe-particle and inter-particle interactions. Experiments were effective for loose deposits on smooth pipe. However, experiments were not effective for cohesive deposits. The cohesive deposits tested actually behaved as loose particles. Also, the water used for testing was of considerably different quality than water for which the pipe coupons had been acclimated. The effects of these water quality differences on deposit properties are unknown.

The Project's field trials involved the most thorough characterization and quantification of flushed contaminants, but were also the most difficult to accurately execute. Relative to the coupons used during pilot testing, the flushing of water mains *in situ* required consideration of several additional variables. This fact, in combination with the invalidation of several data sets, made it difficult to draw sound conclusions about some variables and reduced the confidence in the trends that were identified. Another limitation of the field trials was the inability to see pipe sections before and after each flush. The low velocity removal efficiency results were useful for determining whether a higher velocity removed more material, but offered no information about the amount of material still within the system or the overall cleaning efficiency. The most meaningful field results were based on numerical integral approximations of bulk water data but in each case, only 2 or 3 data points were available.

CONCLUSIONS

This study demonstrated that the factors that govern flushing effectiveness are intricately related to pipe type, water quality, level of treatment, routine hydraulics, and previous pipe cleaning practices. Figure ES.2 illustrates the complexity associated with targeting specific flushing velocities. The top row of boxes represents those variables that can be identified and controlled by a water utility, and are referred to as "control variables". The middle row represents intermediate variables, which are considerably more difficult to evaluate and quantify. Flow arrows are used to define the presence of a significant relationship between sets of variables (boxes) in different rows.

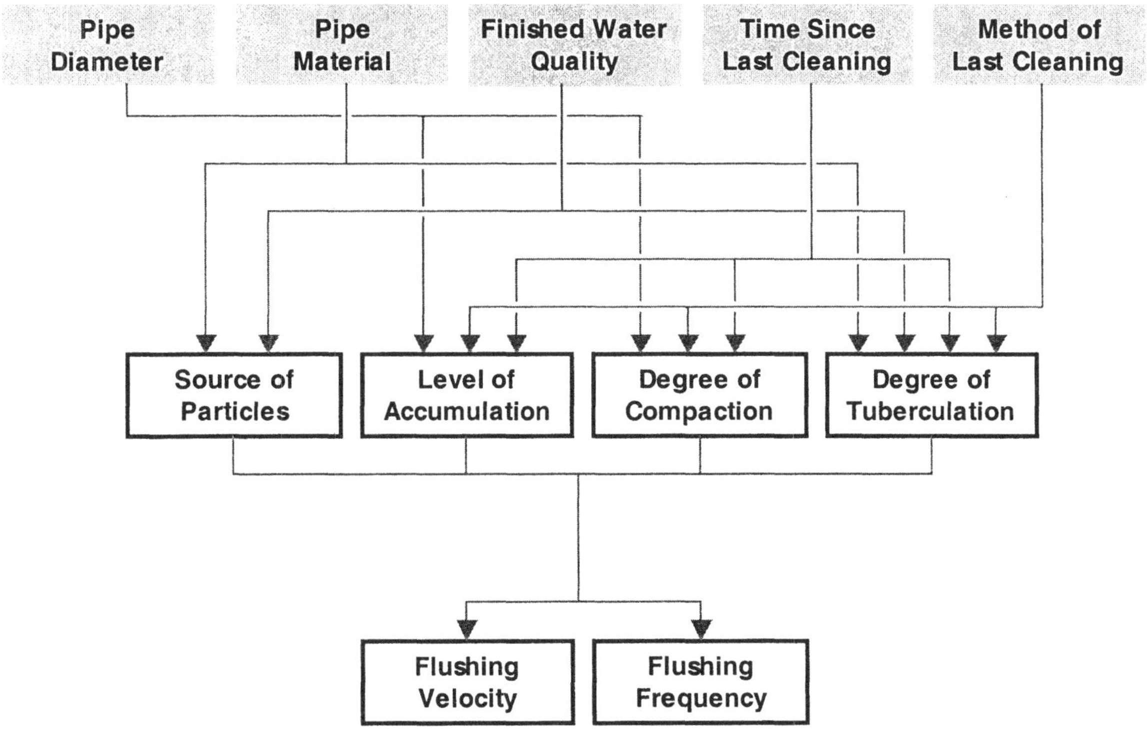

Figure ES.2 Variables impacting flushing effectiveness

It may not be realistic to develop "contaminant-specific" flushing velocities for use within the distribution system. Rather, "site-specific" velocity recommendations can be developed since there are several processes that impact the stability/removability of deposits in distribution mains. This study identified criteria and conditions which can be used to gauge the relative significance of these processes, and hence determine optimal site-specific flushing velocities.

RECOMMENDATIONS

The initial aim of this research project was to develop contaminant-specific flushing velocities. Multiple approaches were used to determine contaminant-specific velocities including CFD modeling, pilot testing, and field testing. Research findings indicated that development of universal contaminant-specific flushing velocities is not warranted. While various distribution system particles may have distinct composition and density characteristics which can be assigned theoretical shear velocities required for scouring and transport, the reality of distribution system conditions requires consideration of particle cohesiveness, adherence to the pipe materials, burial and compaction, site-specific hydraulics, as well as other variables. Thus, findings from this study have been used to develop site-specific flushing velocity guidelines (versus contaminant-specific velocity guidelines) for use by water utilities.

A Flushing Velocity Decision Tree (Figure ES.3) was developed to serve as a tool that allows systems to determine if they are currently flushing at appropriate velocities. The Decision Tree can be used by utilities that want to focus on prevention of customer complaints by removing loose deposits from the system (i.e., low velocity flushing), or by utilities that want to use flushing as a tool for cleaning distribution system piping (i.e., high velocity flushing). Potential negative secondary impacts associated with high velocity flushing were summarized as points of consideration prior to implementation.

Since most systems are comprised of various pipe materials that were installed over many years, there is no single flushing velocity that is optimal on a system-wide basis. Thus, utilities may want to tailor flushing programs to anticipated system conditions so that water quality objectives can be met while minimizing resources used. Velocities ranging from 2 fps (0.6 m/s) to ≥ 5 fps (≥ 1.6 m/s) are recommended, depending on flushing objectives, pipe conditions, historical flushing practices, and routine hydraulics. The concept of "self-cleaning" mains is included. Under these hydraulic conditions, flushing may provide no additional water quality benefit to the utility.

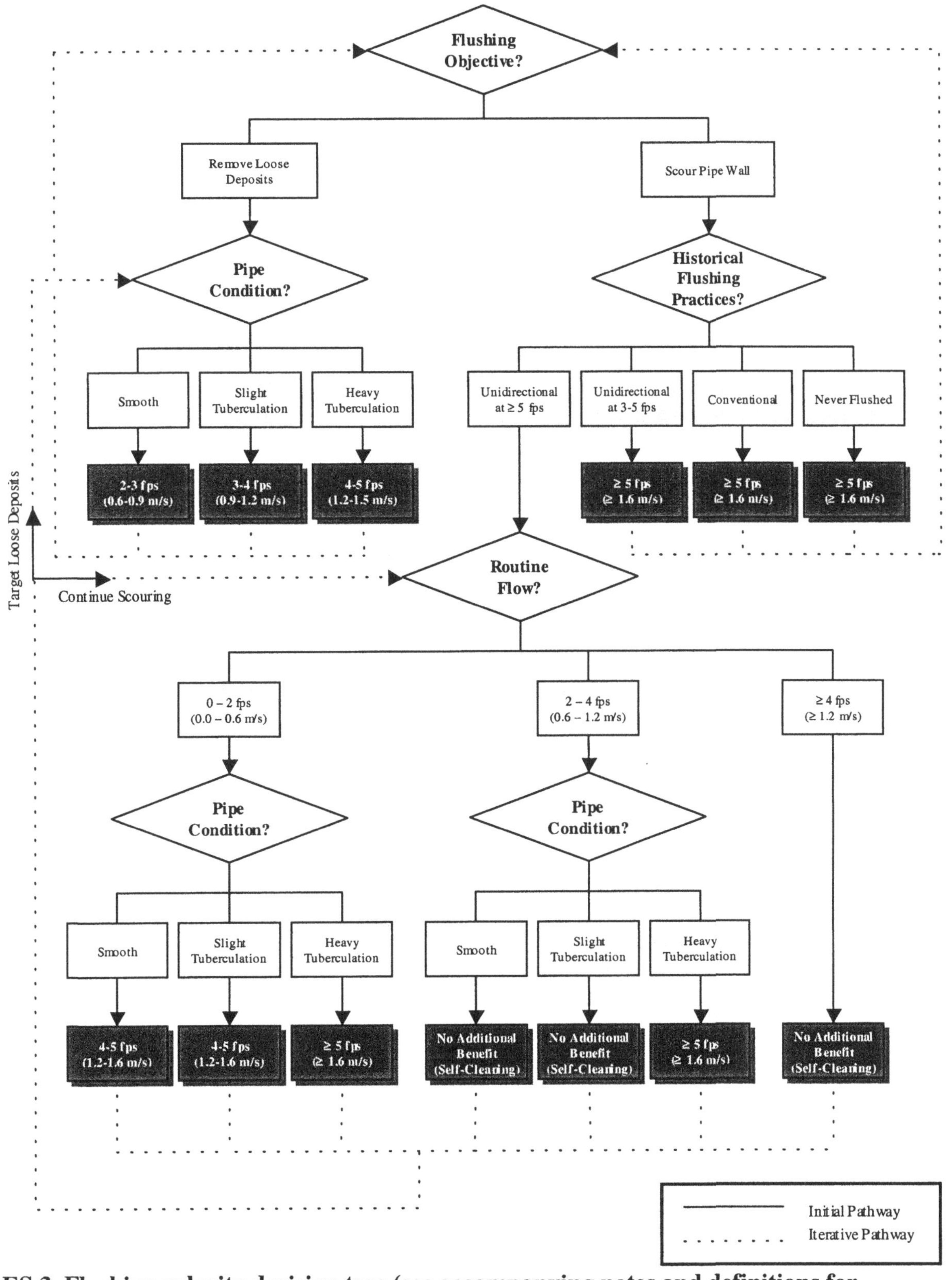

Figure ES.3 Flushing velocity decision tree (see accompanying notes and definitions for using decision tree tool on following page). (continued)

Figure ES.3 (cont.)

NOTES AND DEFINITIONS FOR USING DECISION TREE TOOL

General Note

This decision tree tool is to be used only in the development of an optimized, routine flushing program. It is not intended to address emergency circumstances, such as contamination due to cross connections or regulatory water quality violations.

Selecting Flushing Objective

- The three most common flushing objectives are to: (1) replace stale water, (2) remove loose deposits and (3) scour and clean the pipe surface. Since replacement of stale water is not velocity-specific, this objective is not incorporated in the decision tree.
- Removing loose deposits – this objective may be selected if mains were recently cleaned or rehabilitated (within 4 years) and the goal is to remove additional accumulated material to prevent customer complaints; or if a low velocity flush is needed to protect pipe lining material.
- Scouring pipe wall – this objective may be selected to remove complex, adherent deposits from pipe walls. Prior to implementing a pipe wall scouring flushing program, utilities should consider potential unwanted side effects that can be associated with aggressive flushing (i.e. generation of red water, release of bacteria and opportunistic pathogens into water column, potential for transient low or negative pressures that could allow pathogen intrusion or backflow if cross connections are present).

Estimating Pipe Condition

- Smooth: 0% area reduction. Common examples of smooth pipe are plastic pipe, new AC pipe, newly lined DI pipe, and newly rehabilitated pipe.
- Slight Tuberculation: 0-1% area reduction. Common examples of slightly tuberculated pipe are older AC pipe in corrosive water, older DI pipe, and newer unlined CI pipe.
- Heavy Tuberculation: 1-40% area reduction. Common examples of heavily tuberculated pipe are old, unlined CI and DI pipe and old, lined DI pipe in corrosive water.

Choosing Routine Velocity

Routine velocities are defined here as pipe flow velocities that occur under average day conditions. Hydraulic modeling may be required to evaluate routine velocities.

Flushing Duration

The utilities participating in this study flushed a minimum of one pipe volume during each step. KIWA and Dutch Water Utilities recommend flushing a minimum of three pipe volumes and until a (visual) inspection shows that the water is clear (Slaats 2001). If a field turbidimeter is used, a turbidity target of $\leq$ 5 NTU is recommended.

RECOMMENDATIONS FOR FURTHER RESEARCH

This research study revealed the difficulties with putting science behind the art of flushing. Future research efforts should employ carefully engineered pilot studies. Pipe coupons from actual distribution systems should be utilized in order to accurately simulate real-world particles. Test water should also be representative of the system from which the pipe was exhumed. In order to ascertain whether flushing is an appropriate cleaning technique under specific conditions, mass balance studies should be conducted in conjunction with pilot tests. It is important to note that pilot study results will be most useful to utilities if they can match the conditions within their system to those applied in a specific study. Therefore, research should be conducted to further develop predictive capabilities of the independent variables previously discussed.

A decision tree should be developed to help utilities choose the appropriate flushing frequency based on system specific conditions. Criteria typically used to establish flushing frequency include water quality data; customer complaint records; a set time frame based on scientific literature, a consultant's recommendation or industry standards; water main material, age and condition; and/or flushing records such as time-to-clear and turbidity. The decision tree should also consider constraints to flushing such as water disposal issues, water conservation restrictions, pressure issues, service interruptions, and customer needs.

CHAPTER 1
INTRODUCTION AND APPROACH

INTRODUCTION

Water main flushing is a relatively simple operational activity which involves purging water from the distribution system via a hydrant or blow-off port. Water quality improvements are attained through the displacement of stagnant water and the scouring of accumulated deposits from main surfaces. Historically, many utilities have flushed water mains in response to customer complaints or following maintenance work. Today, many utilities are implementing regular, structured flushing programs as part of a multifaceted strategy to maintain water quality in the distribution system. However, utilities need guidance related to appropriate flushing velocities in order to achieve a balance between water quality improvements, demand for O&M resources, and water conservation. While there are known drawbacks associated with high velocity flushing, including water wastage, environmental disposal, and infrastructure maintenance issues, there is growing concern within the water industry that the use of higher velocities (i.e., $\geq$ 6 fps (1.8m/s)) offers little additional benefit in terms of water quality improvements.

Most often, utilities that practice flushing are interested in removing specific contaminants such as corrosion products, microorganisms associated with biofilms, various chemical scales attached to pipe walls, and suspended or loose material that has accumulated in dead-ends or low-flow portions of the distribution system. The formation and type of contaminant may determine the appropriate flushing velocity associated with scouring, lifting, and transporting the contaminant from the distribution system. Contaminants can occur in different forms in distribution system pipelines, including:

- Material adhered to pipe walls
- Loose material recently settled or generated on the invert of the pipe
- Material on the invert of the pipe that has compacted to form a cohesive deposit
- Particles in suspension

PROJECT OBJECTIVE

No scientific database is available to guide utilities in determining the appropriate flushing velocity that will create a shear force capable of dislodging and transporting specific types of particles, film, or scale. Often, criteria used to decide when flushing is adequate include visual observation of "clear" water coming from the hydrant, a pre-selected time period (sometimes related to calculations on volume of water displaced), the ability of the zone to maintain pressure, or samples collected and analyzed for standard water quality analyses. On-site water quality analyses may be valid indicators of effective flushing but still leave the estimated velocity, duration, and frequency of flushing to guesswork or generalization.

The stated objective of this project was to determine the range of flushing velocities required to dislodge and remove various accumulated materials from water distribution mains.

RESEARCH STRATEGY

The project approach was broken into five tasks: (1) an evaluation of existing information through a literature review, utility survey, and utility case studies; (2) characterization of accumulated materials; (3) pilot studies; (4) field testing; and (5) development of site-specific velocity recommendations. After completing the first task, it became apparent that little published information was available on materials characterization, and that additional field sampling would be needed. Therefore, Tasks 2 (characterization of accumulated materials) and 4 (field testing) were completed concurrently.

Review Existing Information

The objective of this task was to collate and review previous studies relevant to this project from information held in the scientific literature and from utility experiences. Three subtasks included a literature review, utility surveys, and utility case studies.

The literature review was designed to address the following topics:

- The characterization and nature of particles and deposits in distribution systems
- The processes of particle settlement, re-suspension, and transport
- The processes of deposit formation, sloughing, and transport

The utility surveys included 23 North American utilities and 15 utilities in the United Kingdom. Survey participants included only utilities that had established flushing programs. The survey forms and a summary of survey results are included in Appendix A. The specific objective of the utility surveys was to document:

- Utility experiences with targeted flushing programs
- Differences and similarities between United Kingdom and North American distribution systems and flushing practices

The objective of the case studies was to capture information from utilities with long-term flushing programs and databases. Specific data of interest to this project included pre- and post-flushing water quality data, records of pipe ages/conditions, information on targeted particles, and records of flushing operations such as recorded pressures or flushing velocities. The case studies sought to answer specific questions such as whether the utilities used targeted velocities, how they determined acceptable velocities, how they judged the success of flushing, and whether they had correlated pipe velocities to particular materials. After development of several preliminary case studies, it became evident that utilities had not selected velocities based on site-specific conditions.

Characterization of Accumulated Materials

The objective of Task 2 was to develop methods to identify and quantify the material that is to be flushed from distribution pipes. These materials have specific physical and chemical properties that affect the shear velocity and bulk flow velocity required for dislodging and transporting them through piping. The primary goal was to determine the physical characteristics of the particles, as they are related to shear velocity and transport velocity. It was also considered important to characterize the gross chemical or microbiological nature of the particles, and cross-reference these chemical properties with the physical characteristics. This cross-referencing is an important foundation to building a reference database of contaminant characteristics for utilities.

Due to limited scientific literature on materials characterization, the main effort of this task was analysis of particle and water samples collected during the utility field trials (Task 4). To investigate the shear velocities associated with the sloughing of adherent iron corrosion deposits, a pilot scale pipe test rig was equipped with pipe sections with iron tuberculation. Similar studies were conducted on cast iron and PVC pipe sections to evaluate cohesive deposits (GAC fines, flocculated iron particles, biofilm, iron sludge and manganese sludge).

A hydrant sampling and analytical protocol was developed for collecting different types of particles and deposits for characterization under known controlled flow conditions and to collect data on the hydraulic characteristics of particles. Water and sediment samples were collected during both low and high velocity flushing. Sample analyses included observational analysis, particle size, particle density, elemental analysis and organic/inorganic analyses. Further details on sampling and analytical methods are discussed in Chapter 5.

Conduct Pilot Studies

The objectives of Task 3 were to evaluate the impact of different flushing rates on various materials through computational fluid dynamics (CFD) modeling and pipe rig experiments. A three-phase approach was adopted:

- Phase I: CFD modeling was conducted to determine flushing velocities needed to dislodge and/or transport loose materials, and assess the effect of flushing in different piping configurations.
- Phase II: CFD modeling results were verified using a 4-m (13-ft) long experimental test rig.
- Phase III: Flushing rates for cohesive and adherent deposits were studied using a 4-m (13-ft) long experimental test rig containing pipe test sections. The test sections were obtained directly from the utility distribution system or were generated using a 0.8-mile (1.3 km) long test rig which enable suitable material to be developed for testing.

The CFD modeling resulted in detailed information about the movement of loose particles of many sizes, shapes, and specific gravities during steady state and transient flow. Model results yielded a wide range of data on boundary shear velocities required to transport many different particles in a wide range of flow environments. The CFD modeling and pilot study results are discussed in Chapters 3 and 4, respectively. Results from these studies are a key component of utility guidelines discussed in Chapter 7.

Conduct Field Testing

The objective of field-testing was to examine the validity of a contaminant-specific flushing velocity approach in practice. This was achieved by conducting full-scale distribution system flushing at participating water utilities. The participating utilities were selected to represent a range of contaminant types, as shown in Table 1.1.

Table 1.1
Summary of project field trials

Utility	Target deposit or other goal
Portland Water Bureau, OR	Sediment
Cleveland Division of Water, OH	Particulate phosphates
Greater Cincinnati Water Works, OH	Red water
Irvine Ranch Water District, CA	Silt and sediment
Southern Calif. Water Co., CA	Chloraminated surface water
Bellevue Utilities Department, WA	Asbestos cement pipe
South Berwick, ME	Iron and manganese sediment, bedrock fines
Dover Water, NH	Iron and manganese sediment, black water
Anchorage Water and Wastewater Utility, AK	Asbestos cement pipe
Thames Water, UK	Sediment
Yarra Valley Water, Melbourne, Australia	Discolored water, coliform bacteria, unfiltered surface water supply
Newport News, VA	Conversion to chloramine

Develop Site-Specific Velocity Guidelines

The objective of Task 5 was to aggregate all information and data obtained from Tasks 1 through 4 into a format that is readily accessible to utilities needing to develop effective flushing programs. The main product of Task 5 is a flushing velocity decision tree tool that allows the operator to input utility-specific information to obtain a resulting flushing velocity. The decision tree was developed to be iterative. Additional guidelines include:

- A list of contaminants amenable to removal by flushing
- Descriptive characteristics to assist utilities with determination of distribution system conditions
- Simple diagnostic procedures for utilities to identify contaminants in the field
- Guidelines on how to evaluate appropriate flushing velocities in the network

CHAPTER 2
LITERATURE REVIEW AND SURVEY OF EXISTING UTILITY FLUSHING PRACTICES

INTRODUCTION

This chapter provides a review of flushing-related information in the literature and in utility surveys and case studies. The purpose of the project's literature review and utility survey tasks was to aggregate data and information that will assist with the development of a scientific approach to flushing. The chemical and physical processes impacting particles in a pipeline environment include precipitation, sedimentation, compaction, surface scale formation, solubilization, sloughing, re-suspension, and transport. This project focused on the role of flushing with respect to sloughing, re-suspension, and transport processes. The literature review presents a summary of current published findings related to the characterization, deposition, dislodging and transport mechanisms of particles and deposits in distribution systems. Utility survey findings related to the selection and use of specific flushing velocities are discussed in this chapter. Other utility survey findings on flushing programs and practices are presented in Appendix A.

Table 2.1 provides a summary of the literature review findings regarding the type of information available on the characterization of materials collected from distribution systems, or data on shear velocities and transport velocities for specific types of materials.

Table 2.1
Summary of literature research findings –
distribution system pipe deposit/particle related data

Type of Deposit (D) or Particle (P) Identified	Elemental analyses of material?	Adhered to Pipe wall?	Settled in Pipe?	Suspended in Bulk Water?	Entrained in Flow?	Data on Size, Shape, Density, Specific Gravity?	Data on shear or lift velocity?	Data on transport velocity?	Source
D-Magnetite	✓	✓				No	No	No	Evans (1937);Ainsworth, Ridgway & Gwilliam (1979); Benjamin, Sontheimer & Leroy (1996); Smith et al. (1997); Butler & Ison (1966); Rijkens (1967); Weidmann (1970); Clement et al. (1998); Valentine et al. (1998)
D-Geothite	✓	✓				No	No	No	
D-Lepidocrite	✓	✓				No	No	No	
D-Siderite		✓				No	No	No	Smith et al. (1997)
D-Calcite		✓				No	No	No	Smith et al. (1997)
D-Hydrocerrusite	✓	✓				No	No	No	Clement et al. (1997)
Tubercles	✓	✓				No	No	No	Tuovinen et al. (1980); Ridgway and Olson (1981); Clement et al. (1997)
Deposits from Hydrants	✓		✓ - some reported to be settled	✓ - some reported to be suspended		No	No	No	De Rosa (1993)
Deposits from Hydrants	✓		✓ - some reported to be settled	✓ - some reported to be suspended		Size & shape	No	No	Gauthier et al. (1997)
Particles in mains			✓		✓	Size & specific gravity	Yes	Yes	Stephenson (1989)
Iron & manganese rust particles in pipe test rig			✓		✓	Size	Yes	Yes	DeRosa and Miller (1994)

(continued)

Table 2.1 (continued)

Type of Deposit (D) or Particle (P) Identified	Element-al analyses of material?	Adhered to Pipe wall?	Settled in Pipe?	Sus-pended in Bulk Water?	En-trained in Flow?	Data on Size, Shape, Density, Specific Gravity?	Data on shear or lift velocity?	Data on transport velocity?	Source
Iron floc in pipe test rig			✓		✓	Size	Yes	Yes	DeRosa and Miller (1994)
Particles in sewer pipe			✓		✓	No- grouped all sizes	Yes for all sizes as a group	Yes for all sizes as a group	Williams (2000); Nalluri et al. (1994)
Particles in pipe test rig	✓		✓		✓	Size, shape & specific gravity	Yes for all sizes as a group	No	Prince (2000)
Turbidity - iron and loose deposits			✓		✓	No	No	No	Van den Hoven and Vreeburg (1992)
Unanchored crop residue (corn, cotton, pine needles, wheat residues, etc.)			**Settled on surface of sand-bottomed flume*		✓	No	Yes	No	Gilley, Kottwitz, and Wieman (1994)

(continued)

Table 2.1 (continued)

Biofilm or Bacteria	Specific bacteria identified?	Adhered to pipe wall	Settled in pipe	Source
Biofilm	*Arthro-bacter, Flavobac-terium, Moraxella, Acineto-bacter, Bacillus, Pseudo-monas, Alcaligene, Achromo-bacter*	✓		Nagy and Olson (1985)
Biofilm	*Gallionella*	✓		Ridgway and Olson (1981)
Biofilm	*Coliform-Kleibsiella pneumoniae, Klebsiella oxytoca, Entero-bacter cloacae, Entero-bacter aerognenes Escherichia coli*	✓		Camper et al. (1998)
Biofilm	HPC (or not specified)	✓	*Some settled in low flow/stag-nant areas after having sheared off the pipe wall*	McMath et al. (1997); Donlan and Pipes (1988); Kerr et al. (1998); Abernathy and Camper (1998)
Biofilm	*Coliform*, heterotrophs, sulfate reducing bacteria	✓		Clement et al. (1998)
Bacteria in tubercles	*Escherichia coli, C. freundii, E. agglomerans, Klebsiell sp., e. aerogneens, Pseudomonas sp., Edwardiella sp., Gallionella sp.*	✓		LeChevallier, Babcock and Lee (1987)

CHARACTERIZATION OF MATERIALS IN THE DISTRIBUTION SYSTEM

Contaminants in the distribution system typically addressed by flushing include corrosion products, silt, sand, biological scale (biofilm), chemical scale, organic and inorganic particulates, and pipe lining material. As discussed by Walski (1991), these materials can degrade aesthetic water quality and threaten public health protection because of their ability to reduce disinfectant residuals and release toxins or pathogenic bacteria. These materials can exist in several different states depending on local hydraulic and water quality conditions:

- Particles in suspension
- Loose material settled or generated on the invert of the pipe
- Loose material that has been layered or compacted to form a cohesive deposit
- Deposits and tubercles adhered to pipe walls

The terminology applied to material collected from distribution mains is rarely consistent in the literature. For example, inorganic material adhered to the pipe wall has been referred to as deposits, tubercles, encrustations, scales or sediment.

Adhered Deposits

The majority of studies on the characterization of deposits in distribution systems have focused on the examination of deposits adhered to the pipe surface. These deposits have attracted research interest due to their involvement in the passivation of the corrosion process. As described in the following paragraphs, research efforts have been focused on describing the structure, chemical composition, and mechanisms of formation of pipe wall deposits. There are relatively few studies that address the adherence and size of pipe wall deposits.

John Baylis and U.R. Evans initiated studies on the nature of pipe wall deposits. Baylis (1926) examined a large number of corrosion tubercles in the Baltimore water supply system and postulated that the formation of a tubercle on an iron pipe was "...not by chance, but by a well-applied force." He described the different parts and processes in a tubercle. This included the formation of an impervious membrane which was described as a crystalline compound of iron with magnetic properties. According to Baylis (1926), the thickness of this layer was dependent on the status of the corrosion process, with dormant tubercles having the thickest membrane layers. Evans (1937) referred to stratified layers of different iron oxides making up a corrosion deposit. These oxides included magnetite (Fe_3O_4) and either goethite (α-FeOOH) or lepidocrocite (γ-FeOOH), depending on the rate of oxidation. These early studies introduced the concept of the pipe wall deposit being a structured, active entity, with stratified corrosion products dependent on the different water chemistries and the developmental stage of the corrosion process. Subsequent studies have focused on characterizing the mineralogical structure of pipe wall deposits.

Figure 2.1 illustrates the typical structure of a corrosion deposit. Kölle and Rösch (1980) described a bottom layer of base metal covered by a layer to which the corrosion process penetrated. This layer was referred to as the topotactic layer and included a mixture of graphitized and original iron material. Above this layer was the epitactic layer, which is where the iron oxides accumulated to form a tubercle that protruded into the bore of the pipe. The boundary of the tubercle was marked by the occurrence of a hard, brittle layer composed primarily of magnetite, with some goethite. This was referred to as the "shell-like" layer. Above this shell was a layer of loose, porous deposit composed of goethite and calcite. This layer was in contact with the water. Table 2.2 provides a summary of minerals commonly found in corrosion deposits and illustrates the diversity and complexity of deposits adhered to pipe walls. Figures 2.2 and 2.3 show other simplified versions of the structure of a typical corrosion deposit as described by Ainsworth, Ridgway and Gwilliam (1979) and Benjamin, Sontheimer and Leroy (1996), respectively. Like Kölle and Rösch (1980), an enveloping "shell-like" layer that enclosed interior corrosion products and was covered by exterior products was described. According to Benjamin, Sontheimer and Leroy (1996), the enveloping layer was an electrical conductor because of the magnetite content. This enabled it to act as a cathode in a sandwich-structured electrochemical cell, with the metallic surface acting as the anode. Many other authors, including Butler and Ison (1966), Rijkens (1967), and Weidmann (1970) have discussed similar structures.

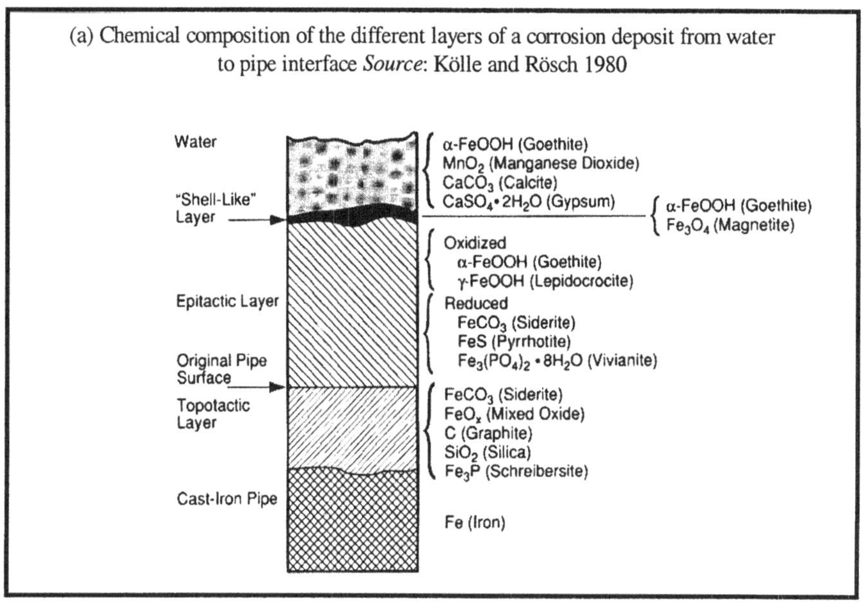

(a) Chemical composition of the different layers of a corrosion deposit from water to pipe interface *Source*: Kölle and Rösch 1980

Figure 2.1 Profile of a corrosion deposit (Kölle and Rösch 1980)

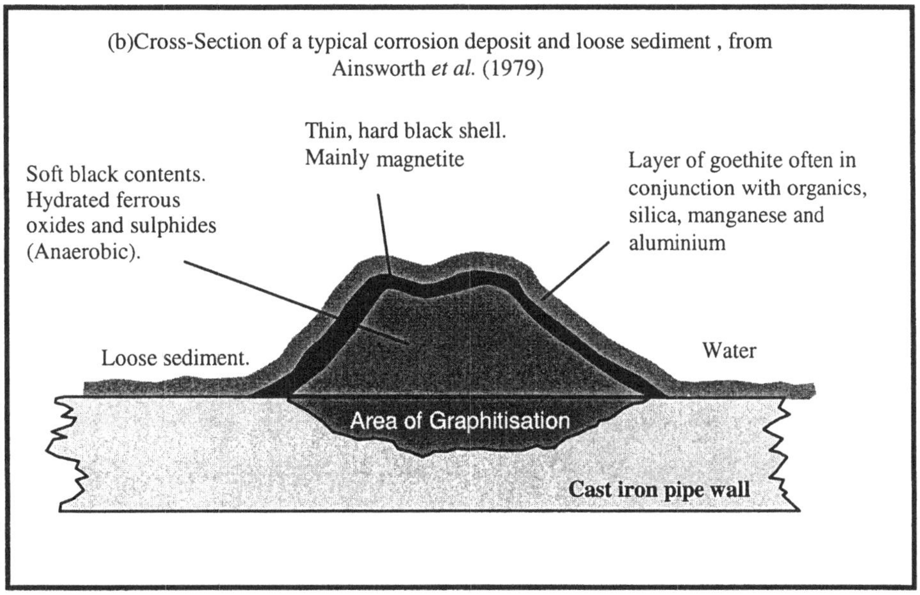

Figure 2.2 Profile of a corrosion deposit (Ainsworth, Ridgway and Gwilliam 1979)

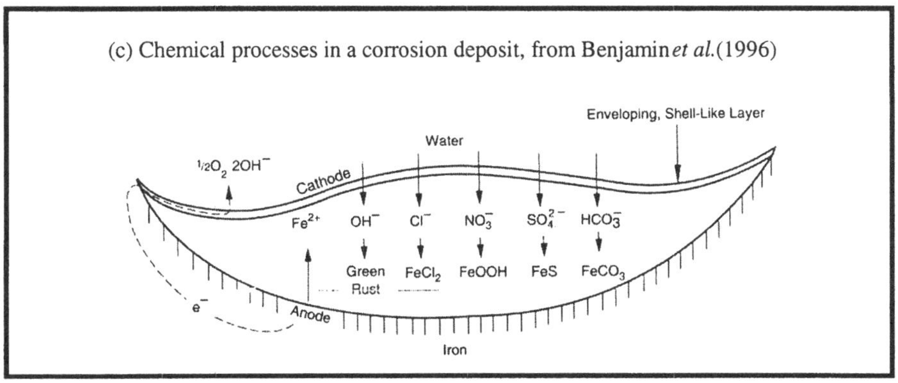

Figure 2.3 Chemical processes in a corrosion deposit (Benjamin, Sontheimer, and Leroy 1996) Reprinted from *Internal Corrosion of Water Distribution Systems, Co-operative Research Report, 2nd Edition,* by permission.

Table 2.2
Minerals commonly found in corrosion deposits

Mineral name	Formula	JCPDS No.
Haematite	Fe_2O_3	13-534
Maghemite	a-$Fe_2 0_3$	15-615
Magnetite	Fe_3O_4	19-629
Goethite	α-FeOOH	29-713
Lepidocrocite	γ-FeOOH	8-98
Wustite	FeO	6-615
-	$Fe(OH)_2$	13-89
-	$Fe(OH)_3$	22-346
Pyrrhotite	FeS	17-200
Gypsum	$CaSO_4.2H_2O$	6-46
Viviante	$Fe_3(PO_4)_2.8H_2O$	3-70
Mackinawite	FeS	15-37
-	$FeSO_4.4H_2O$	16-699
Monohydrocalcite	$CaCO_3.H_2O$	17-528
Vaterite	$CaCO_3$	13-192
Anhydrite	$CaS0_4$	6-226
Aragonite	$CaCO_3$	5-453
Calcite	$CaCO_3$	5-586
Wollastonite	$CaSiO_3$	10-487
Foshagite	$Ca_4Si_3O_9(OH)_2$	11-94
Whitlockite	$Ca_3(PO_4)_2$	9-169
Hydroxyapatite	$Ca_5(PO_4)_3(OH)$	9-432
Siderite	$FeCO_3$	12-531

More recently, Smith et al. (1997) studied the chemical composition of corrosion deposits in Thames Water's distribution system in southeastern England. The research revealed that pipe wall deposits were very similar in composition. Deposits were composed primarily of inorganic material, with a mixture of iron oxides (magnetite), iron oxyhydroxides (goethite and lepidocrocite) and iron and calcium carbonates (siderite and calcite) forming the main constituents. Goethite and magnetite were the most ubiquitous minerals.

Clement et al. (1998) collected pipe samples from several U.S. cities to characterize the scale and biofilm observed on the pipes. The pipe samples included cast iron, lead, and galvanized steel. The cast iron pipes had once been cement-lined, but at the time of sampling, the lining had worn off. The pipes varied in age from 15 to greater than 100 years. The authors used visual inspection, scanning electron microscopy (SEM), energy dispersive spectroscopy, and x-ray diffraction (XRD) analyses. Table 2.3 provides a summary of the results. The dominant elements observed on the cast iron pipes were iron, calcium, silica, and sulfur. The dominant elements observed on the lead pipes were lead, aluminum, and silica. The dominant elements observed on the galvanized pipe were iron, aluminum, and silica. The mineral goethite

(α-FeOOH) was found on all the cast iron pipes and on one galvanized pipe sample, while magnetite (Fe_3O_4) was found on nearly all of the cast iron pipes, a lead pipe, and a galvanized pipe. The mineral hydrocerussite ($Pb_3(CO_3)_2(OH)_2$) was found on the lead pipes.

Table 2.3
Summary of scale analysis results

Type of pipe	Major compounds	Major elements	Comments
Cast iron	Goethite, magnetite	Fe, Ca, Si, S	Silica from cement-lined pipe.
Lead	Hydrocerrusite	Pb, Si, Al	Al from past use of aluminum-sulfate coagulant.
Galvanized	Magnetite, goethite, lepidocrocite	Fe, Si, Al	

Source: Data from Clement et al. (1998)

An important observation emanating from the studies of pipe wall deposits is the different adhesive properties of deposits. Terms such as "loose, porous" or "dense, protective" have been used to describe deposits in distribution systems. Researchers have concentrated mainly on characterizing deposits with adherent, protective properties and identifying the water quality conditions that might explain their formation. "Protective deposits that form on pipe surfaces may be a mixture of corrosion products that depend on the type of metal that is corroding and the composition of the water solution." (Schock 1990).

Sontheimer, Kölle and Snoeyink (1981) addressed the role of siderite ($FeCO_3$) in the formation of protective deposits. They discussed the importance of the formation of relatively insoluble solids containing reduced iron compounds, such as siderite, in the development of dense, protective deposits. If conditions prevented their formation, then loose, porous deposits resulted. These latter deposits were easily sloughed off, leading to discoloration and increased corrosion rates. The critical reaction involved the ferrous ions produced by the initial oxidation of the metal surface. This reaction was thought to be strongly dependent on the water quality at the deposit/water interface. The authors admit that the siderite model contains many oversimplifications that may hinder its application to a live distribution system. Despite these points, this model is important because it provided a mechanism to explain field observations of different deposit structures and properties.

Biofilms and Microorganisms

The mechanisms of biofilm formation and detachment in the distribution system are of great interest to water utilities as these biological scales may contribute to the potential threat of waterborne disease. Biofilm formation may also accelerate the corrosion of water mains and cause aesthetic water quality problems.

Characterization of biofilms and identification of the factors controlling biofilm regrowth have gained increased attention in the research community over the last decade. Biofilms have been found in most natural and industrial aquatic systems. They consist of a matrix of cells and cellular products attached to a substratum, where they grow, reproduce, and produce extracellular polymer layers (Peyton 1996; Dombay, Piriou, and Kiene 1999; Emde, Smith and Facey 1992). Tuovinen and Nurmiaho (1979) present evidence that groundwater can be the origin of distribution system bacteria by showing that iron-rich groundwater supports the development of organisms tentatively identified as *Leptothrix* and *Gallionella.*

The body of research literature about biofilm regrowth is significant, and the focus for this project was specific to the topics listed below:

- The types and densities of organisms commonly present in distribution system biofilms.
- The characteristics of their habitat and attachment to the pipes.
- The tendency and conditions for their detachment from the pipe (due to mortality, flushing, etc.).
- The occurrence and behavior of suspended biological scale.

Microorganisms on Pipe Surfaces

Donlan and Pipes (1988) designed a study to quantify bacterial populations that colonize freshly exposed cast iron pipe surfaces. Their goal was to correlate water quality characteristics with accumulated microorganisms. The study estimated population densities of attached microorganisms using the HPC procedure, and then verified the counts with a direct microscopic procedure. Population densities after 28 days of attachment to cast iron pipe surfaces ranged from $10^{4.3}$ – $10^{7.57}$ CFU/cm^2. Abernathy and Camper (1998) conducted a bacterial regrowth study using a laboratory reactor equipped with mild steel surfaces and operated to simulate the physical conditions found in distribution systems. The reactors operated for a period of about 16 weeks, after which time the biofilm density was measured using the HPC procedure. Population densities were similar to those found by Donlan and Pipes (1988), ranging from approximately $5.01 \times 10^5 - 1.58 \times 10^6$ CFU/cm^2.

Nagy and Olson (1985) found that HPC densities on various types of pipe surfaces ranged between 10 bacteria per 100 cm^2 and 4.7×10^6 bacteria per 100 cm^2. They reported that *Arthrobacter* were predominant on the pipe surface, and that other bacteria associated with biofilms included *Flavobacterium, Moraxella, Acinetobacter, Bacillus, Pseudomonas, Alcaligenes*, and *Achromobacter.* They observed a correlation between bacterial density on the pipe surface and the age of the pipe, estimating that for each year of service, a ten-fold bacteria increase occurred.

Kerr et al. (1998) also recognized that pipe material affects biofilm development. They analyzed heterotrophic colony development on cast iron (CI), Thermanox, medium density polyethylene (MDPE), and unplasticized polyvinyl chloride pipes (uPVC). The CI pipe exhibited the shortest population doubling time and the highest steady-state biofilm development, followed by Thermanox, MDPE, and uPVC pipe. The biofilm heterotroph counts on CI pipe were 10-100 times higher than on the MDPE and uPVC pipes.

Microorganisms Within Tubercles

Several researchers have shown that coliform bacteria are present within corrosion tubercles. LeChevallier, Babcock and Lee (1987) identified *Escherichia coli* (fecal coliform positive), *C. freundii*, and *E. agglomerans* (greater than 160 bacteria per gram) within tubercles collected from bituminous-coated, cement-lined cast iron pipe. Emde, Smith and Facey (1992) found that total coliforms were present in greater numbers in corrosion deposits than in the untreated water supply. The authors examined corrosion tubercles from exhumed pipes and found 22 types of bacteria, including *E. coli, Klebsiella sp., E. aerogenens, Pseudomonas sp., Edwarsiella sp.,* and *Gallionella sp.*

Allen, Taylor, and Geldreich (1980) used SEM to examine tubercle and encrustation samples from seven utilities across the U.S. The physical characteristics of the seven tubercles were similar in that they were all observed to have hard but porous surfaces. In many cases, microorganisms such as actinomycetes, algae, and diatoms were seen at or near the surface.

Emde, Smith and Facey (1992) investigated microbial-influenced corrosion associated with low temperature water (8-20° C). They observed that corrosion tubercles most commonly supported iron-reducing bacteria, including *Bacillus* sp., *Pseudomonas* sp., *Psuedomonas fluorescens*, *P. aeruginosa*, *Citrobacter* sp., *Klebsiella* sp. and *Edwardsiella* sp., *Leptothrix, sphaerotilus*, and *Gallionella.* However, in a study of corrosion tubercles collected from bituminous-coated cast iron pipe, Tuovinen et al. (1980) observed no iron-reducing bacteria at a temperature of 28 °C. They did report the observation of sulfate-reducing and nitrate-reducing bacteria, as well as sulfur-oxidizing, nitrite-oxidizing, and ammonia-oxidizing bacteria.

Loose Deposits

There have been few studies published involving large-scale investigations of the nature of loose deposits in distribution systems. Of the few studies conducted, the findings have been focused on the characterization of loose deposits in terms of their chemical composition and their occurrence in the distribution system. There is little data on the physical characteristics of deposits, such as density and size. Table 2.4 summarizes the research findings.

Table 2.4
Microbial characteristics of loose deposits collected from various distribution systems

Sample collection method	Major constituents found	Bacterial description	Bacterial density	Source
Flushed water collected from hydrant	Fe: 40-70% Al: 1-6% Ca: 2-13% Mn: 0-4% Organics: 14-24%	Coliforms sampled in 4 systems, total bacteria sampled in 1 system, HPC sampled in 2 systems	Total coliform: 3 positive samples of 258 samples; Total bacterial counts: 10^{10} cells/g HPC: 10^{6} to 10^{7} CFU/g	Carrière et al. 2002
Hydrant net (142 μm)	Fe: 13 - 45% Al: 2 - 45% P: 0.1 - 18% Ca: 0.5 - 17% Mn: 0.2 - 6%	N.S.	N.S.	DeRosa 1993
Hydrant net (100 μm)	Fe: 12.7% Ca: 4.4% Al: 5.3% Mn: 2% Volatile Solids: 18.9%	Diatoms, gastropods, microscopic crustaceans, nematodes, ciliated and flagellated protozoa, HPC	N.S. except for HPC which were found at 2.5 x 10^{8} CFU/14 d-gram	Gauthier et al. 1997
Reservoir sediment	Fe: 38.8% Ca: 3.6% Al: 2.1% Mn: 0.4% Volatile solids: 12.9%	N.S.	7.2 x 10^{7} CFU/14 d-gram	Gauthier et al. 1997
Water collected from hydrant and filtered (0.2 μm)	N.S.	Rod-shaped bacteria	Bacteria present on 17% of the 10-50 μm sized particles collected	Ridgway and Olson 1981
Flushed water collected from hydrant	N.S.	Filamentous and stalked bacteria spread across particle surface; Also colonized coliform on particles by adding nutrients to water	N.S.	Herson et al. 1991
Flushed water collected from hydrant	N.S.	Total bacterial counts	1 x 10^{11} bacteria/gram	Barbeau et al. 1999
Hydrant net (300 μm)	N.S.	Coliform	0.5 bacteria/gram	LeChevallier et al. 1987

N.S. = not specified by author.

Carrière et al. (2002) conducted unidirectional flushing in four Canadian distribution systems and evaluated the accumulated loose deposits as a function of time, pipe material and water characteristics. The quantity and nature of the deposits vary greatly among the various networks and sampling campaigns. The principal factors affecting deposit accumulations are the accumulation period, the pipe material, and the aggresivity of the water.

DeRosa (1993) conducted a survey encompassing 71 supply zones in the United Kingdom. Distribution system deposits were collected with hydrant nets (aperture size 142 microns) and bulk water samples were taken at intervals over the flush. Elemental analyses were carried out on both the water and sediment samples from the hydrants, as well as samples taken from customer's taps. The field surveys showed that there was a large variation in the quantity of deposits in water mains within supply zones. Even mains of similar size and type and in close proximity showed large variations in quantities of loose deposit present. It was suggested that local features (e.g.

hydraulics) of the water mains had a significant effect on particle deposition and accumulation. Relatively more deposits were found in small diameter mains (≤ 4 inches) at the distal ends of distribution systems, an observation attributable to the occurrence of low flow regimes at these sites. However, large diameter mains (≥ 6 inches) were not devoid of sediment. Statistical analysis indicated that lowland reservoirs and river sources (which tend to contain more fine sediment from overland runoff) were associated with more distribution deposits than upland reservoirs and groundwater well sources. The compositional analysis of the deposits collected in hydrant nets produced a list of seventeen elements. The predominant elements were iron (13-45% of the sample), aluminum (2-45%), phosphorus (0.1-18%), calcium (0.5-17%), and manganese (0.2- 6%). DeRosa (1993) did not quantify particle size or density of the deposits.

Gauthier et al. (1997) conducted a similar loose deposit survey in the distribution system of the Communauté Urbaine du Grand Nancy, France. This system serves surface water to a population of 300,000, and contains cast iron, cement-lined ductile iron, PVC, and polyethylene pipes. Loose deposit samples were collected from hydrants and service reservoirs. The study attempted to define the nature of the deposits in the distribution system and to evaluate differences between the nature of deposits and their location in the network. Despite the fact that the system was supplied by a single water source, the appearance and composition of loose deposits varied considerably between different mains and finished water reservoirs. The authors identified the following organisms in the loose deposits: diatoms, gastropods, microscopic crustaceans, nematodes, and ciliated and flagellated protozoa. Heterotrophic plate counts were used to quantify culturable bacteria. Bacterial densities in pipe deposits averaged 2.5 x 10^8 CFU/14 d-gram. The colors of deposit ranged from reddish to yellow-brown to dark brown. The texture of deposits ranged from grains to floc-sized particles. This variation was in agreement with the findings of DeRosa (1993) and could be an indication of the importance of local system features in determining the occurrence and nature of deposits in the network. Compositional analysis showed the main constituents were iron, volatile solids (principally organic matter), insoluble material, calcium and aluminum. The organic matter content was much smaller than the mineral component. No data on particle size or density was recorded.

In their study of water quality impacts of dead-end flushing, Barbeau et al. (1999) correlated total suspended solids (TSS) and total bacterial counts during flushing, suggesting that loose deposits were colonized by bacteria. Using direct total bacterial counts, they estimated a bacteria colonization of 10^{11} bacteria per gram of suspended solids. The authors compared these findings to those of Gauthier et al. (1999) and Emde, Oberoi and Smith (1995), where bacterial densities had been calculated based on HPC measurements. In both cases, the bacterial densities were found to be lower (by 400 and 2,000 times, respectively) than these direct count measurements. They suggested that HPC underestimates bacterial density by only enumerating culturable bacteria, and/or by enumerating small particles that are colonized by numerous bacteria as a single colony.

LeChevallier et al. (1987) examined sediment (primarily sand) collected from a New Jersey distribution system where water sampling indicated the presence of coliform. They identified 0.5 coliform bacteria per gram and HPC levels of 8.4 x 10^6 bacteria per gram.

Marshall (2000) characterized deposits collected from hydrant flushing in four UK water utilities. Chemical analysis showed that all sediment samples contained high concentrations of iron and manganese. Particle size analysis showed particles were in the range 1 to 162 microns. Ackers, Brandt and Powell (2001) summarized and studied the characteristics of loose material using pipe test rigs. Ackers, Brandt and Powell (2001) concluded that sediment in water may consist of corrosion products, pipe lining, treatment breakthrough, animal and biomatter and/or calcium deposits. The specific gravity of sediment was stated to be theoretically between 1.0 and 7.0. The authors recommend a characteristic specific gravity value of 3.0 to represent inorganic sediment. Particles between 2 μm and 20 mm were observed in the field trials. The authors concluded "...there is a dearth of published data on the physical and chemical characteristics of sediment particles within distribution networks. In particular there is a lack of data for the size, specific gravity and cohesive properties of sediment particles."

Suspended Particles

There is minimal published information addressing particles in the bulk water column. Examination of the nature of particles causing discolored water has been rare, perhaps due to difficulties analyzing what is often very small amounts of material or because the interpretation of the source of the particles is very difficult.

A recent project by Slaat (2001) characterized the processes that govern the generation of discolored water. Discolored water incidents are the result of sediment resuspension in the distribution network. The project included development work on a technique called the resuspension potential method for predicting discolored water events in the distribution network. In this method, a standardized and reproducible disturbance is introduced in the water flow (flow increase of 0.35 m/s (1.15 fps) on top of normal flow) and maintained for 15 minutes. The water turbidity is continuously measured before, during and after the disturbance. The change in turbidity during the disturbance is a measure of the quantity of sediment resuspension. Studies conducted on various sediment types in a smooth pipe with no biofilm showed that the velocity at which particles first move is approximately 0.10 m/s (0.43 fps) for light sediment and 0.15 m/s (0.65 fps) for heavier sediment. The light particles are all in motion at 0.15 m/s (0.65 fps) and the heavier ones at 0.25 m/s (1.1 fps). No relationship was observed between the pipe diameter and the velocity at which all particles are in motion.

Gauthier et al. (1998) collected suspended particles from a distribution system by filtering tens to hundreds of liters of water through 5-μm membranes or glass fiber filters. Microscopic examination revealed a wide range of particle shapes, structure, and colors. Suspended particles were generally less than 20-μm in diameter. Gauthier et al. (1998) noted the predominance of matter in suspended particles compared to inorganic deposits found on pipe walls. This was reported to be due to the lower densities (not quantified) of the organic material, which facilitated suspension and transport.

De Rosa's (1993) work was one of the few studies where a detailed chemical analysis of suspended particles had been carried out. Filtered particles (pore size 0.45-μm) were analyzed for their inorganic constituents. The results were compared to data collected from deposit samples at the same locations. There was a wider range of elements present in the deposits. The filtrate samples contained sodium, calcium, magnesium, phosphorous, and potassium present as the main constituents, with trace concentrations of aluminum, iron, zinc and manganese. The deposits contained these elements as well as chromium, cadmium, copper, lead, mercury, and nickel.

Walker (1983) presented a classification system of the different particle types in discolored water. This system included drawings of typical distribution particles together with their probable source. These drawings are reproduced in Figure 2.4.

TYPE	APPEARANCE (Not to scale)	PROBABLE SOURCE
Asellus FRASSE		Asellus infestation of main or service reservoir.
'TEA LEAF' Mn-RICH SCALE		Iron main local to complaint.
'TEA LEAF' Pb-RICH SCALE		Lead piping. Usually accompanied by white lead carbonate.
WHITE CRYSTALLINE		Lead carbonate from corrosion of lead pipe. Often follows sloughing of scale.
DENDRITIC		Crystals grow on lining of main carrying water with soluble Mn content. Usually remote.
BLACK AMORPHOUS		Probably broken-up dendritic particles (but may be bitumen or activated carbon if metal stains negative)
FLOCCULAR		Low density ferric hydroxide but may have organic & bacterial content. Local or remote corrosion. May contain aluminium.
'HUMIC' FRAGMENTS		May be derived from decay of vegetable matter in sediments.

Source: Walker (1983)

Figure 2.4 Particles commonly found in distribution systems

Smith et al. (1998) used a pilot-scale pipe test rig to study the origin of red water in distribution mains. In this study, the authors found a statistically significant positive relationship between turbidity, color and total iron, and were able to use the relationships in pilot testing to predict red water generation. They tested both continuous flow pipelines and dead-end pipelines, each containing the same amount of unlined cast iron. Red water generation was observed in the dead-end, but not in the continuous flow pipe.

PARTICLE SETTLEMENT, RE-SUSPENSION AND TRANSPORT

The suspension of particles in the bulk water column can be the result of two mechanisms: (1) metal uptake, or (2) entrainment of pipe surface deposits. The former was investigated as part of the AwwaRF study *Development of Red Water Control Strategies*, while the latter is the focus of this study. Hydraulic entrainment involves the physical suspension of loose deposits induced by flow changes which affect the deposit-water interface. Entrainment is the result of a force imbalance on the deposit. Once the particle is in suspension, transport is determined by buoyancy, weight, drag in the flow direction, and drag in the vertical direction.

Most studies on particle entrainment have focused on studying the velocities and tractive stresses required to initiate the movement and lifting of deposits. Studies on particle entrainment and particle transport in suspension, as well as sediment transport modeling, have been reviewed. For a detailed review of sediment transport theory, see Ackers, Brandt and Powell (2001).

Particle Entrainment

Stephenson (1989) determined the velocities required to suspend and transport coal 200-µm particles. The research considered the impact of both particle density and pipe diameter. The paper was based on early work by Durand and Condolios (1952), who studied transport characteristics of heterogeneous mixtures (mostly coal) as a high concentration slurry. Stephenson's data, which are provided in Table 2.5, exhibit a strong dependence on pipe diameter. For example, a two-fold increase in velocity is required to remove particles from an 8-inch (200-mm) main compared to a 2-inch (50-mm) main. These flow rates are currently widely used in the UK water industry to determine flushing velocities.

Table 2.5
Flow required to suspend and transport 0.2 mm coal particles

Pipe diameter in (mm)	Velocity fps (m/s)		Flow rate gpm (L/s)	
	SG = 3.0	SG = 1.5	SG = 3.0	SG = 1.5
2 (50)	4.3 (1.3)	2.3 (0.7)	42.8 (2.7)	23.7 (1.5)
3 (75)	5.2 (1.6)	2.6 (0.8)	114.1 (7.2)	60.2 (3.8)
4 (100)	5.9 (1.8)	2.9 (0.9)	237.8 (15.0)	120.5 (7.6)
6 (150)	7.2 (2.2)	3.6 (1.1)	650 (41.0)	317 (20.0)
8 (200)	8.5 (2.6)	4.3 (1.3)	1,316 (83.0)	666 (42.0)

Source: Stephenson (1989) SG = Specific Gravity

DeRosa and Miller (1994) studied the lifting of particles under experimental conditions using a 4-inch diameter pipe test rig. Three types of particles were studied: rust, manganese/iron mix, and iron floc. The particles were size-graded by sieving prior to seeding. The pipe rig was operated at flow rates of 48, 79, and 111 gpm (3, 5, and 7 L/s). The percentage of deposits recovered in a hydrant net secured at the end of the pipe was calculated. At a flow rate of 79 gpm or 2 ft/s (5 L/s, 0.64 m/s), the following recoveries were observed: 97% of the fine deposits (i.e., smaller than 0.15 mm); 18% of the medium-sized deposits (0.15-1.18 mm); and 0% of the coarse deposits (1.18-3.35 mm). At a flow rate of 111 gpm or 2.7 ft/s (7 L/s, 0.88 m/s), 22% of the medium-sized deposits and 6% of the coarse deposits were recovered.

An early rule of thumb used in the design of sewer pipelines was to maintain a minimum velocity of 2 fps (0.6 m/s) to ensure silt-free conditions. Williams (2000) reported that at velocities above about 1.6 fps (0.5 m/s) most loose sediment in mains is suspended in the flow and is likely to be transported through the mains. Nalluri and Dabrowski (1994) used self-cleaning velocities and minimum shear stress approaches to investigate flow depth as an important factor in the build-up of deposits in sewers. Self-cleaning velocities are a function of pipe size (hydraulic radius), flow depth, sediment size, and pipe roughness. Findings indicate that for flow depths ranging from half-full to full, there is a self-cleaning velocity for each pipe size. However, if the pipe is less than half full, deposition occurs. An important point to remember, however, is that accumulated sediment in the pipe changes flow hydraulics, which can lead to either decreases in the bed shear stress (and increased accumulation), or localized increased velocities and shear stress (and possible resuspension of solids).

Prince, McManus and Goulter (2000) examined the lifting velocities of particles from Melbourne, Australia's water distribution system. The particles used for testing were collected from a service reservoir during cleaning. Particle characterization was conducted for the following: specific gravity (hydrometer test), size (sieve analysis), particle shape (SEM), and elemental composition (EDXA). EDXA analysis revealed that the primary elements in the particles were clay and iron. The sources of these constituents were believed to be soil in the catchment and the corrosion of pipe, respectively. The specific gravity of particles was found to be 3.07. The particles were predominantly fine silt, smaller than 6-μm in diameter. The results of laboratory tests performed to determine particle scour velocities are summarized below.

- First visual movement at 0.20 fps (0.06 m/s)
- Peeling movement of light particles at 0.36 fps (0.11 m/s)
- Most sediment removed at 0.72 fps (0.22 m/s)
- All sediment removed at 1.1 fps (0.34 m/s)

Based on review of current sediment transport models, Ackers, Brandt and Powell (2001) suggest new velocity recommendations for effective flushing. These recommendations vary between 2.0 fps (0.6 m/s) for 2-inch (50-mm) pipes with roughness coefficient of 1.5 mm and 5.6 fps (1.7 m/s) for 24-inch (600-mm) pipes with roughness coefficient of 0.01mm.

In a different approach to particle movement, Gilley, Kottwitz, and Wieman (1994) studied the hydraulic conditions required to initiate movement of unanchored crop residue materials (e.g., corn, cotton, sorghum, soybean, pine needles, sunflower, wheat) in a sand-bottom flume. The researchers measured discharge rate and flow velocity required to initiate residue movement. Testing revealed that bouyancy forces significantly influenced incipient motion of large diameter crop residues, while drag forces were more critical in the movement of the smaller residues. Perhaps one of the most interesting findings was that as the residue rates increased (and the residues themselves overlapped), so did the stability of the residue layer. A more stable residue layer required increased critical shear stress to initiate motion.

Van den Hoven and Vreeburg (1992) and Van den Hoven et al. (1994) investigated the different causes of discolored water using continuous turbidity monitors at various locations in distribution systems in the Netherlands. They hypothesized that there were two primary mechanisms for the creation of turbid water: (1) the release of iron from the mains and (2) the entrainment of loose deposits in the main. These two causes were shown to produce different diurnal turbidity patterns. In the Amsterdam network, where iron release problems predominate, turbidity peaks were associated with periods of low consumption. Here, the turbidity rose during the night, until midnight, and decreased in the early morning when water consumption increased. According to the paper, this relationship between turbidity and residence time indicated that iron release was the cause of discolored water. Spot sampling confirmed the presence of iron containing particles in the system.

In another distribution system of the South-Kennemerland Water Supply Company (WLZK), the turbidity patterns were almost the reverse with periods of high consumption coinciding with turbidity peaks. This turbidity pattern was thought to be caused by the entrainment of loose deposits or, as Van den Hoven and Vreeburg (1992) described, the "whirling up of sediment present in the network as a result of drastic changes in magnitude, or, direction of the water flow". Pipe sections removed from the distribution network confirmed this. Corrosion products and tuberculation were absent and a thin bituminous coating was still intact.

Transport of Particles in Suspension

The settling velocity of particles or sediment is one of the most important parameters in the description of particle transport. Traditionally, the settling velocity has been determined in a settling column such as the Owen tube.

DeRosa (1993) and DeRosa and Miller (1994) studied the settling characteristics of deposits from different parts of distribution systems. DeRosa (1993) made the distinction between samples of "pure" iron oxide of high density (not quantified) that settle quickly and samples of organic floc of lower density which settle slowly (up to 4 days to clear). These observations conform to the theoretical settling rates of particles in water, as expressed by Stokes Law (Sears, Zemansky, and Young 1988). Stokes Law (equation 2-1) dictates that for a given viscosity of water, the smaller the particle, the smaller the terminal velocity and, therefore, the greater the settling time.

$$U = \frac{2r^2}{9\eta}(\rho_p - \rho_f) \qquad (2\text{-}1)$$

where: U = Terminal velocity
r = Particle radius
η = Fluid viscosity
ρ_p = Particle density
ρ_f = Fluid density

At present, the determination of settling velocity of discrete particles involves tedious and repeated trial-and-error calculations. Bhargava and Rajagopal (1992) developed nomograms for the direct evaluation of the settling velocities of various discrete particle diameters having a specific gravity in the range of 1.001-2.65 and temperature in the range of 5-40°C. Using these nomograms, an integrated expression was evolved based on the polynomial fitting concept to predict the settling velocity of discrete particles in relation to the particle diameter, specific gravity of the particle and kinematic viscosity (or temperature) of the suspending liquid.

Larsen (2000) emphasized the difficulties of measuring the settling velocities of fine particles due to their non-constant state (e.g., flocculation). According to Larson (2000), settling velocities of fine floc have a range of apparent settling velocities depending on the local conditions. The use of re-circulating settling columns, as developed by Rasmussen (1997), was recommended for establishing the apparent settling velocities of fine particles.

Woodward et al. (1995) investigated the movement of particles in a 0.8-mile (1.3-km) pipe test rig. Results from a step flow trial validate the assumption of plug flow (for turbulent conditions), since the particle profile is propagated through the rig without dispersion. The number of particles increases for a short period of time sharply with flow step increase. The explanation for this phenomenon is the existence of "hot spots", in which loose deposits preferentially accumulate. These deposits are lifted into suspension with a step increase of the flow rate.

SEDIMENT TRANSPORT MODELS

Much of the development of sediment transport models in the UK has been initiated to tackle the problem of sediment accumulation in sewers. In the UK, many sewers are laid to shallow gradients resulting in recurring sediment deposition problems. The loss in carrying capacity due to these deposits may contribute to hydraulic overloading problems and ultimately pollution of the local watercourses. Therefore many UK publications on sediment transport address dirty water networks. Examples of papers which detail the theory of sediment transport and derivation of model input parameters are given below.

Ackers, Brandt and Powell (2001) review sediment transport theory and its relation to clean water networks. The authors reviewed several sediment transport equations and found that the most applicable to clean water distribution systems were those in papers by May (1994) and Macke (1982). The Ackers-White equation, regarded as one of the most accurate equations for sediment transport, was considered to be inapplicable to clean water distribution networks since the sediment load and potential deposition depths are too small.

Alasdair and Ashley (1999) used a deposition factor, which increases with critical shear stress, to provide an estimate of the sediment build up rate within drainage pipes. Babaeyan-Koopaei, Butler and Davies (1999) carried out an experimental program to investigate the advection behavior of solids. The critical shear stress was found to be directly proportional to the Reynolds number of the particles. From the data of Xu (1996), they found that the shear stress associated with 50% deposition is well- described using equation 2-2.

$$\tau = A \times \ln(Re) - B \qquad (2\text{-}2)$$

Where: τ = Critical shear stress
A = Constant
B = Constant
Re = Reynolds number

The critical shear stress has been applied in the transportation of sediment in the alluvial streams. It is found that the critical stress is directly proportional to the size of the particle (Garde and Ranga 1985).

Crabtree (1989) detailed results from an investigation of the fundamental hydraulic, rheological (deformation of sediment in pipe caused by flow) and biochemical properties of sewer sediments. This research was commissioned as part of the development of deterministic models for the settlement, consolidation and re-entrainment of sewer sediments in relation to the hydraulic regime within a sewer. The particle characterization results are not relevant but the approach and thinking are of great interest to flushing clean water distribution systems. Crabtree (1989) stated the limitation of current design criteria for sediment transport in sewers was that sediments are assumed to be 'clean' non-cohesive individual particles. However, real sewer sediments contain significant amounts of organic matter which give cohesion to the deposit, requiring a higher critical yield stress and higher shear stress to initiate entrainment. Crabtree (1989) reported that the previous work which had shown the sediment cohesion and consolidation processes are time-dependent. A two-phase system for the transport, deposition and erosion of sewer sediments was postulated:

(1) Transport and deposition of 'clean' sediments using conventional non-cohesive sediment transport theory related to particle settling velocity and bed shear stress conditions during dry weather flow; and
(2) Post-depositional changes (consolidation, biological action) which lead to the development of a 'dirty' cohesive sediment bed which require the use of cohesive sediment transport theory to predict entrainment.

Crabtree (1989) carried out determinations of the rheological characteristics (critical yield stress) of sediments samples using applied stress rheometry. All samples exhibited a degree of cohesion and a range of critical yield stresses from 0.6-1.9 psf (25-88 N/m^2) was recorded. Shear stress used for the design of self-cleaning pipes was 0.04-0.09 psf (2-4 N/m^2).

Nalluri and Alvarez (1992) found that even a low level of cohesion can greatly increase the value of shear stress required to initiate sediment movement and that, once entrained, the cohesive sediments tended to behave as a non-cohesive sediment.

Skipworth, Tait and Saul (1996) carried out a series of laboratory experiments on transport rates of 100-mesh ground olivestone. The experiments produced evidence that the measured suspended sediment transport rate of cohesive-like materials exposed to a variable flow rate, is dependant both on the values of the final flow rate and on the rate of the initial temporal change of flow rate, and is not solely dependent on the steady-state flow variables. According to Skipworth, Tait and Saul (1996), the behavior of the cohesive-like sediment bed can be split into two phases – an initial part in which the suspended sediment transport rate rises dramatically followed by a phase in which there is an equally rapid decline in the transport rate to a stage in which the transport rate is almost constant. The rate of movement in the first phase appears to be related to the rate of flow rate increase, while the second phase is partially dependent on bed shear stress.

Sakakibara (1996) carried out experiments on flushing trunk mains sewers in which he studied particle characteristics, tractive force and critical particle size. The results are not relevant here but the approach is of interest. The particles were a mixture of silts and sands, the organic material having a specific gravity of 2.49 and the inorganic material having a specific gravity of 2.59. An equation is presented to calculate the tractive force and critical particle diameter from the results of water velocity and depth. The specific gravity of particles was assumed to be 2.65 in the equation. The tractive force was calculated as 0.006 psf (0.3 N/m^2) and the critical particle size was predicted as 0.36 mm.

More recently Zug et al. (1998) described the development of a sediment transport model for sewers utilizing the Velianov total load model in which sediment transport is modeled by the Velianov theory. This requires calculation of settling velocity and particle density. Mineral particles or 'coarse' particles were given a settling velocity of 2.5 ft/min (45 m/hr) and a relative density of 2.65. Organic particles or fine particles (i.e., median diameter of 50 microns) were given a settling velocity of 0.3 ft/min (5.4 m/hr) and a relative density of 2.00. These input parameters were verified by field measurements and trials at sites in France.

There are only a few papers in the literature that address the development of sediment models for clean water distribution systems. For example, Maier (1998) used results from step flow increase trials on a 0.8-mile (1.3 km) pipe test rig to develop a total particle count model for distribution systems. Maier's (1998) particle count model describes transient biofilm detachment in terms of particle counts. The model incorporates the concept of a local shear-off function that allows the development of a linear model to describe shear-off from biofilms that comes about due to a flow increase. A biofilm shedding profile was determined which provides a novel way of testing distribution systems for critical areas in terms of particle shear-off. Maier (1998) found that transient flow increases have a decisive impact on the amount of particles in the distribution system over a period of several days. In addition, transient detachment may account for the removal of significant amounts of biofilm. According to Maier (1998), previous publications are chiefly concerned with medium- or long-term variability of biofilm thickness due to shedding (or shear-off). Thus, the time dependence is generally assumed to be linear (i.e., a rate per time unit is found) or slowly varying. Under controlled conditions, Maier (1998) recorded sharp peaks in particle counts as flow was increased. These peaks propagated almost in tact through the rig. After a step down in flow, the total particle counts were observed to decrease slowly. In Maiers' test rig trials, organic particles (suspended bacteria) formed the predominant portion of counted particles. Most particles were in the 2-5 micron range.

Kerneïs, Martin and Deguin (1996) produced a model of particle transport used to determine zones of particle deposition in a part of a distribution system in East Sussex, England (South East Water Company distribution system). According to Kerneïs, Martin and Deguin (1996), the concentration of suspended particles was determined by a combination of three forces; drag force, gravity force, and buoyant force. By incorporating an equation of these forces into a hydraulic model of the network, areas where hydraulic conditions promoted deposition were identified. Hydrant flushing was then carried out at these points to assess the quantity of deposits. If there was no significant quantity of deposits in the pipes where depositing conditions were predicted, then the upstream mains were assumed to be in good condition. Iron and manganese analyses were also carried out on the deposits and on the source water to determine the source of iron. Using this strategy, areas of iron deposition and iron corrosion were identified. The results showed that large quantities of iron deposits were found in non-iron based mains, suggesting that the predominant mechanism to explain the occurrence of loose deposits in the mains was entrainment rather than local iron release from corroded mains. The source of the deposits was stated as being from iron corrosion not the input water.

The model developed by Kerneïs, Martin and Deguin (1996) illustrated the importance of understanding the hydraulics of a network when interpreting the occurrence of particles and subsequent discoloration. Loose deposits were not found in areas of high flow velocity. Also, the effect of network configuration on hydraulics was shown, with most loose deposits being found at dead ends rather than in looped zones in the system where dead ends are absent. The study illustrates the value of incorporating hydraulic information into the design of flushing programs.

Sethi et al. (1999) studied particle entrainment and deposition in distribution systems by using the results of studies on pneumatic transport of solids, and gas phase particle deposition and re-entrainment; these mechanisms were applied by similitude to waterborne particles. The influences of pipe geometry and Reynolds number on the pipe wall shear stress were also examined. Since successful deposition of particles is likely to be controlled variably by viscosity, electrical double-layer interactions, and van der Waals forces, they believe that particle transport "is a necessary but not a sufficient condition for deposition". The inertial effects involved in particle entrainment were discussed in terms of particle size. For example, for particles smaller than 30 microns in size, they found that adhesion forces needed to be exceeded for lift off. For particles of size 30-200 microns, entrainment "stress" is independent of particle size. For particles greater than 200 microns in size, gravitational forces dominate and the critical shear force increases with particle size.

Francis and Minton (1984) also considered the hydrodynamic fluid forces acting on a particle. Drag forces act in the direction opposite of flow, while lift forces act normal to the flow direction. Saffman (1965) studied lift forces exerted on small spheres. According to Saffman (1965), in the boundary layer, lift forces are produced on the particle due to the mean shear in the fluid. The force is proportional to the square of particle diameter and square root of velocity gradient. Another force is the thermophoretic force, which is proportional to the temperature gradient (Talbot et al. 1980). Particles are moved towards cold areas and repulsed from hot areas.

There are three types of sediment transport models commercially available in the UK: WATSED, LiQUAL, and Piccolo. All have a sedimentation feature and each uses a different empirical relationship between particle behavior and flow parameters. The relationships have been produced for rivers, estuaries and solids transport. Although some work has been done to verify the usefulness of these models, much more research is needed. In the meantime, these models can produce useful indications of the parts of the network where sediment can collect.

BIOFILM SLOUGHING

Biofilm can be removed from the pipe surface by the erosive action of flowing water, a phenomenon referred to as sloughing. McMath et al. (1997) conducted sloughing studies using both biofilm and settled bacteria. The experiments consisted of a series of flow increases to determine the effect on the numbers and nature of particles in water. A sustained peak of clumped bacteria was seen for all the flow changes indicating that biofilm was being sloughed off the pipe walls even after several increases in flow. Ten-fold increases in flow resulted in up to 60% of the bacteria in the water phase being present in microcolonies or clumps. Based on the 4-inch (100-mm) pipe diameter, the flow rates tested correspond to flow velocities of 0.0033 fps (0.001 m/s) and 0.037 fps (0.01 m/s), respectively.

The increase in the number of bacteria was most noticeable in the first flow increase. In this case, the biofilm had been allowed to develop on the pipe over a period of 6 weeks with minimal bulk flow and in contact with water absent of disinfectant residual. The high increase in numbers of bacteria present were attributed to the re-entrainment of settled deposits and the shedding of areas of the biofilm which had developed during the disinfectant-free regrowth period. Baseline conditions were not achieved until after 48 hours of flushing, and clumps continued to be removed even after a number of successive flow increases. This work is of interest because it provides an indication of the time that elevated bacteria numbers can be expected when flow is increased in low flowing parts of the distribution systems. Further, the continual shedding of biofilm during successive flow increases indicates that biofilm regrowth can occur rapidly.

Donlan and Pipes (1988) analyzed microbial densities on coupons placed in the Pennsylvania Suburban Water Company's system (cast iron pipe). Their work verified the occurrence of sloughing by showing a linear correlation between maximum flow velocity and HPC density at six sampling sites. For reference, pipe velocities ranged from 0.04 fps (0.012 m/s) to 2.49 fps (0.76 m/s). As the velocity in the pipe increased, the population of HPC attached to the coupon decreased, citing the possibility that either an increased flux of disinfectant or physical removal (shear) were responsible for the decrease.

Kerr et al. (1998) also identified a significant, positive correlation between flow rate and biofilm viable counts in a laboratory model system. The authors did not report flow rate or velocities achieved in these tests. Three pipe types were used: CI, MDPE, and PVC. Cast iron pipes supported the highest steady state bacteria population.

OVERVIEW OF FLUSHING APPROACHES

In this section, basic information is provided on common flushing methods as related to this research effort. Depending on the utility-specific water quality objectives and hydraulic/maintenance considerations, one or more flushing approaches may be appropriate. Two basic flushing approaches are addressed in this document – unidirectional flushing and conventional flushing. Newer flushing methods practiced in Europe are also discussed.

Unidirectional Flushing

Unidirectional Flushing (UDF) consists of isolating a particular pipe section or loop, typically through closing appropriate valves, and exercising the hydrants in an organized, sequential manner. It is carefully engineered with consideration to flushing crew size, duration of flushing, equipment availability, and location of water sources such as water treatment plant, storage tanks, booster stations, etc. The term unidirectional is often associated with a velocity of 6 fps (1.8 m/s) (Antoun, Tyson and Hiltebrand 1997), which is thought to be adequate for biofilm removal and the removal of corrosion products and other debris attached to the pipe walls. However, the concept of isolating pipe segments and flushing in a sequential manner from the source to periphery (i.e., unidirectional) can be practiced at lower velocities to achieve different water quality goals such as removing loose deposits, restoring chlorine residual, reducing tastes and odors, and restoring flows and pressures. Some benefits of unidirectional flushing include:

- Potential for water savings compared to conventional flushing
- The generation of written procedures that allow reduced management time during flushing operations
- Long-term water quality improvements
- Standardized procedures that provide uniform basis of comparison with future flushing events
- Improved valve maintenance and identification exercises

Conventional Flushing

Conventional flushing is the approach currently used by most utilities. It is usually implemented with little, if any, pre-design and engineering, and significantly less planning than unidirectional flushing. Conventional flushing consists of opening hydrants in a specific area of the distribution system until pre-selected water quality criteria are met. These criteria could include restoring disinfectant residual, reduction/elimination of color, reduction in turbidity, etc. It is important to note that valve isolation is *not* part of conventional flushing. Consequently, flushing velocities are not maximized as water to the hydrant often flows from several mains with the resultant velocity in each individual main remaining low. Because water used to flush a particular main during conventional flushing may not originate from a segment that has already been flushed, the cleaning efficiency is not maximized. Further, sediment and deposits may simply be shifted from one main to another, rather than being purged from the system. The primary water quality improvements that can be achieved through conventional flushing may include:

- Restoration of disinfectant residual
- Expulsion of some of the poor water quality in a specified area of system

If these are the primary water quality goals within a specific portion of the system, conventional flushing may be adequate. However, according to Antoun, Tyson and Hiltebrand (1997), the following problems have been experienced as a result of conventional flushing:

- Increased customer complaints during and immediately after implementation
- Water wastage
- Minimal, short-lived water quality benefits
- Potential for increased coliform occurrences

Comprehensive and Spot Flushing

Each flushing approach can be implemented on a comprehensive or spot basis. The term "comprehensive" indicates that the flushing campaign addresses the entire system. "Spot" flushing is typically used to target an acute problem in an isolated area of the system, or in an area where chronic water quality problems have led to repeated customer complaints. It is typically implemented at a higher frequency than comprehensive flushing. Since spot flushing focuses more on replacing the bulk water (vs. cleaning the pipe), water quality benefits are typically short term.

New Flushing Methods

Pulsating Flushing

"The cleaning principle in pulsating flushing is to loosen and resuspend sediment particles in the pipe by accelerating the water's velocity. The flow is increased within a short period according to a fixed pulsating pattern. The increased velocity results in a brief but sharp increase in wall shear stress. The acceleration is achieved by the controlled, repeated opening and closing of a valve at the end of the section being flushed. In between the periods of increased velocity, flushing takes place at a constant basic velocity, which is sufficient to carry along the loosened particles." (Slaats 2001)

Flowjet Technique

A flushing technique called "FLOWJET" has been developed by the WRc in the United Kingdom and has been proven to remove films and deposits using turbulent flushing with less water than other methods (Slaats 2001). The technique uses a tethered device to provide a blockage to the normal flow in the pipe. The differential head across the device is used to produce high velocity flow through small apertures in the device. This high velocity flow is directed at the pipe walls by careful alignment of the apertures. The jets formed are able to remove deposits from the pipe walls and lift sediment lying in the pipe. Field trials have demonstrated that the flow-jetting method can be used successfully to remove loose deposits, sediments, and adherent deposits.

FLUSHING EXPERIENCES AND OPTIMIZATION

Examples of papers where flushing is cited as a tool to improve water quality in North American and European distribution systems are summarized below. Related information from the project utility surveys and case studies is also discussed.

Many researchers and utility managers have recommended optimization of flushing programs as part of the solution to water quality improvement and maintenance (Emde, Oberoi, and Smith 1997; Emde, Oberoi, and Smith 1995; Smith, Smith and Milosky 1996; Pattison 1980; Barbeau et al. 1999; Antoun, Tyson, and Hiltebrand 1997).

KIWA and the Dutch water companies have recently drawn up some basic principles for implementing flushing programs (Slaats 2001):

1. Water flushing is the most effective technique for cleaning mains.
2. When flushing, the flushing rate of the water must be at least 1.5 m/s (5 fps).
3. The volume of flushing water must be at least 3 times the volume of water in the pipe, until a (visual) inspection shows that the water is clean.
4. The work is performed on the basis of a clean water front.
5. Making measurements, in addition to carrying out the cleaning activities, increases effectiveness.

van Lieverloo et al. (1998) reported on how flushing, air scouring and pigging were used in trials as techniques to control invertebrates. Mains were flushed to give a water flow velocity of 3.3 fps (1.0 m/s) in the distribution mains. The flushing program was devised according to recommendations of an earlier study by Vreeburg (1996). Cleaning of mains was advocated as a temporary solution to be used in situations that require an immediate reduction of invertebrate numbers. Cleaning the entire distribution system after optimization of treatment was recommended to be most effective.

Flushing was one of the measures adopted by Zurich Water Supply, Switzerland, to minimize regrowth of microorganisms following implementation of the supply of non-chlorinated water. In 1993, a full-scale practice of distributing water without adding any chemicals at the outlets of the treatment works was implemented. In order to control bacterial regrowth, various measures were adopted. This included flushing the network to avoid the occurrence of stagnant water. According to Klein and Forster (1998), the hydrants in Zurich, Switzerland are flushed twice a year. No details are given of the methodology used. However, Klein and Forster (1998) report that on occasions when HPC up to 100-300 CFU/mL were recorded, the affected pipe sections were flushed intensively, generating flows of 1.6 fps (0.5 m/s) in the distribution main.

Vibert, Gayaud and Amyot (1991) describe the deployment of a mobile unit for the detection and evaluation of sediments during flushing. Hydrants were flushed at flow rates of 60 m^3/h (264 gpm) and the flushed water was fed into the mobile sediment assessment unit which incorporated on-line turbidity monitoring and quantified suspended solids collected via a hydrocyclone. The clean thresholds adopted were turbidity of 4 NTU and suspended solids equal to 20 mg/liter. The mobile unit was reported to result in optimized flushing programs and reductions in costs.

Though there is a general awareness that flushing improves water quality and reduces complaints, there exists no database or guidance on targeting specific pipe deposits or contaminants to improve water quality and maintain optimal conditions within the system. Water quality modeling can been used to predict the effects of flushing, as was demonstrated by Walski and Draus (1996) in their 1994 study of the Fort Monmouth, New Jersey distribution system where the USEPA's water quality model EPANET was used to predict chlorine residual and turbidity levels. Their findings indicated that flushing alone would not solve the problem of low chlorine residuals in the system during the summer months.

In their nationwide survey on distribution system flushing programs, Chadderton, Christensen and Henry-Unrath (1992) correlated a lack of record keeping with the inability to evaluate flushing effectiveness. They found the most common objective of flushing programs was cited as response to customer complaints. Less than half of the utilities collected flushing water samples for laboratory analyses, and most depend on 'time to clear' as a measure of flushing completion. The most common laboratory analyses cited were chlorine, turbidity, and bacteria.

Emde, Oberoi, and Smith (1995 and 1997) estimated that unidirectional flushing removed almost twice the amount of biomass as did conventional flushing, and recognized that unidirectional flushing allowed higher velocities, and therefore greater boundary shear at the pipe wall. Smith, Smith and Milosky (1996) recommended unidirectional targeted flushing in the Port Charlotte, Florida distribution system to ameliorate and prevent nitrification in a distribution system using chloramines. Unidirectional flushing was cited as a way to actually move sediment out of the system instead of merely transporting it from place to place. Barbeau et al. (1999) evaluated the benefits of "spot flushing" of dead-ends in a portion of Montreal's distribution system. These researchers found that the effectiveness of spot flushing varied between two locations monitored; in fact, the two locations exhibited inconsistent trends in turbidity and total iron. Their experience indicates that even when dead-ends receive the same source water, contain the same pipe materials, and have the same hydraulic configurations, water quality variability resulting from spot flushing can be expected.

Antoun, Tyson, and Hiltebrand (1997) present a case study describing unidirectional flushing of 70 miles (113 km) of pipe at the U.S. Army Garrison Fort Belvoir in Fairfax County, Virginia that eliminated repeated occurrences of positive coliform samples. These authors recommend flushing at 6 fps (1.8 m/s) or greater to successfully transport biofilm out of the system. A recent report by Joseph and Pimblett (2000) discusses the success of unidirectional flushing of all pipes 12-inches and smaller in the Virginia Beach, Virginia distribution system. These authors describe a significant improvement in the efficiency of the Virginia Beach flushing program after switching from conventional to unidirectional flushing. Customer complaints were reduced due to the shorter period of time required for cleaning each pipe segment.

COMPARISON OF UK AND U.S. DISTRIBUTION SYSTEMS

As part of an UKWIR study (Cardew et al. 2000), the differences between U.S. and UK distribution systems characteristics were reviewed. The findings of that report are summarized here.

In the UK, the typical per capita consumption is approximately 34-45 gal/person/day (130-170 L/person/day). In the U.S., the demand is much higher at approximately 120 gal/person/day (454 L/person/day) and varies considerably in different regions of the country. Also, the design of the system must consider fire-fighting demands in addition to normal domestic and industrial water use.

The piping in a distribution system includes transmission lines, distribution mains, and service lines. Transmission lines are long pipelines connecting supply sources to treatment facilities, and treatment facilities to the distribution system. Distribution mains convey water from the transmission mains to customer service lines.

An estimated 880,000 miles (1,416,200 km) of piping are in service in the United States (Kirmeyer, Richards and Smith 1994). According to a survey of the ten water supply and sewerage service utilities in England and Wales (Latimer 1996), a water company has, on average, some 14,900 miles (24,000 km) of water mains. These are composed of a diverse range of materials of varying ages of service. However, most will be composed of iron, asbestos cement and plastic materials.

According to Economic and Engineering Services and Kennedy/Jenks/Chilton (1989), the types of pipe in common use for transmission and distribution systems throughout North America are steel, galvanized, cast iron, ductile iron, reinforced concrete, asbestos cement, plastic, and wood. An estimated 13,200 miles (21,240 km) of new piping are being installed annually (Kirmeyer, Richards and Smith 1994). The predominant forms of new piping are cement-lined ductile iron (47.7%), PVC (38.7%), and concrete pressure (12.5%).

The types of water mains in use today are summarized in Table 2.6. The summary is based on a survey of 998 utilities in the U.S. (AWWA and AWWARF 1992) and a 1996 survey in the UK (Jack 1996).

Table 2.6
Water main materials used in the U.S. and UK

Type of pipe	Percent of total miles	
	U.S.	UK
Cast iron with cement lining	25.8	19
Cast iron without cement lining	22.2	19
Coal tar lined iron	0.2	28
Ductile iron with cement lining	16.7	**
Ductile iron without cement lining	2.4	**
Asbestos cement with vinyl lining	1.9	**
Asbestos cement without vinyl lining	13.2	17
PVC (in UK includes polyethylene and PVC)	9.2	14
Steel	4.0	**
Concrete pressure	2.2	**
Lead	0.1	**
Other	2.1	3
Total	100	100

** Unknown
Sources: AWWA and AWWARF 1992; Jack 1996

The survey shows that the predominant material is iron, comprising over 60% of U.S. and UK distribution systems. It must be noted that the above percentage figures are approximations since many water companies lack comprehensive pipe material databases.

There are significant differences in the range of pipe sizes in UK and North American distribution systems. A typical UK distribution system may contain a range of pipe sizes from 12-inch (300-mm) feeder mains down to 3-inch (75-mm) and often 2-inch (50-mm) pipes at the end of the network. In the U.S., pipe sizing is designed to accommodate fire flows and there are few pipes of less than 6-inch (150-mm) in the networks. Most mains are 6, 8 or 12-inch (150, 200, 300-mm). Pipes 36-inches (900-mm) and larger are frequently used as major transmission lines in large water systems.

UTILITY SURVEY RESULTS

The purpose of the project survey was to collect information about utility flushing experiences that reflect specific observations and/or data collection with respect to targeted flushing velocities, or targeting specific pipe materials/deposits for removal. Background and general information gathered from the utility surveys are presented in Appendix A. Of 36 surveys sent to U.S. and Canadian utilities, 23 (64%) were completed and returned. A utility flushing survey was already being carried out in UK as part of another research project; these data were reported here rather than repeating the exercise. The only difference between the UK and U.S. surveys was that the UK survey did not request background information on distribution system characteristics, sediment sources and maintenance practices.

Targeted Flushing Velocities

Fourteen of 23 North American surveyed utilities have established specific flushing velocities; two of these 14 utilities used specific velocities to target unique contaminants for particular water quality objectives. That reported target for both utilities was 2.5 fps (0.8 m/s) for particulate phosphate and a low velocity for removing sediment. Seven of the 14 utilities using specific flushing velocities cited velocities in the range of 5-7 fps (1.5-2.1 m/s); one utility used a 2.5 fps (0.8 m/s) velocity. The method of identifying the proper velocity to meet objectives was determined by various means including field testing/prior studies (6 of 14 utilities), trial and error (5 of 14 utilities), research literature (4 of 14 utilities), and consultant recommendation (2 of 14 utilities). Three of the 14 reporting utilities varied flushing velocity by pipe type, material, age or condition. Thirteen of 14 utilities reported areas of the distribution system where the specified velocities could not be achieved. In these areas, 3 of the 14 utilities have attempted alternatives to flushing including pigging, swabbing or longer flushing times at lower velocities. Eight of 13 responding UK utilities reported they did not have preset velocities or flushing times to target.

CHAPTER 3
COMPUTATIONAL FLUID DYNAMICS

INTRODUCTION

Computational fluid dynamics (CFD) is a mathematical based technique applied to simulate fluid flow around structures. It is a computer-based method for the solution of the fundamental equations of fluid dynamics: conservation of mass, momentum, and energy. CFD was initially applied to problem solving in the aerospace, automotive, and power generation industries. In the water industry, CFD attracts interest as it offers a reliable and non-intrusive means to investigate the performance of many process units. In most cases, CFD is employed when the conventional measurement and/or engineering design rules meet difficulties. The technique is also employed as a preliminary tool to evaluate design options.

CFD has been used in this Project to improve understanding of the flow patterns and particle movement in water mains. CFD is used to simulate fluid flow through smooth, slightly tuberculated, and severely tuberculated pipes in order to calculate the velocities required to lift various particles from the pipe floor into the flow. The output from the CFD modeling is the calculation of the velocities required to lift particles, of certain sizes and densities, in pipes of different diameters and degrees of tuberculation. CFD has been applied to determine lifting of loose particles only. The lifting of adherent or cohesive deposits was studied using experimental pipe rigs and is discussed in Chapter 4.

MODELING OBJECTIVES

CFD modeling was conducted to meet the following objectives:

- To calculate the required flow velocities for lifting and removing particles of various sizes and densities, in pipes of different diameters and tuberculation states.
- To provide improved understanding of the effect of pipe diameter and condition, and particle size and density, upon flushing efficiency.

METHODOLOGY

The work comprised the following stages:

(1) Build and solve 3 CFD models (smooth, slightly tuberculated, and severely tuberculated) for each of three pipe sizes: 4-inch (100-mm), 6-inch (150-mm), and 8-inch (200-mm). Representations of the 3 states of tuberculation are shown in Figure 3.1.
(2) Develop a methodology to calculate the lifting velocities for various particles in the above pipes.
(3) Calculate the fractional pipe area cleaned for the three pipe conditions, for each of the
(4) three pipe diameters, and for particles of different sizes and densities.
(5) Present results graphically and provide raw data as a database.

The detailed methodology for building the CFD model (stages 1 through 3) is given in Appendix B. Stage 4 is presented and discussed in this section.

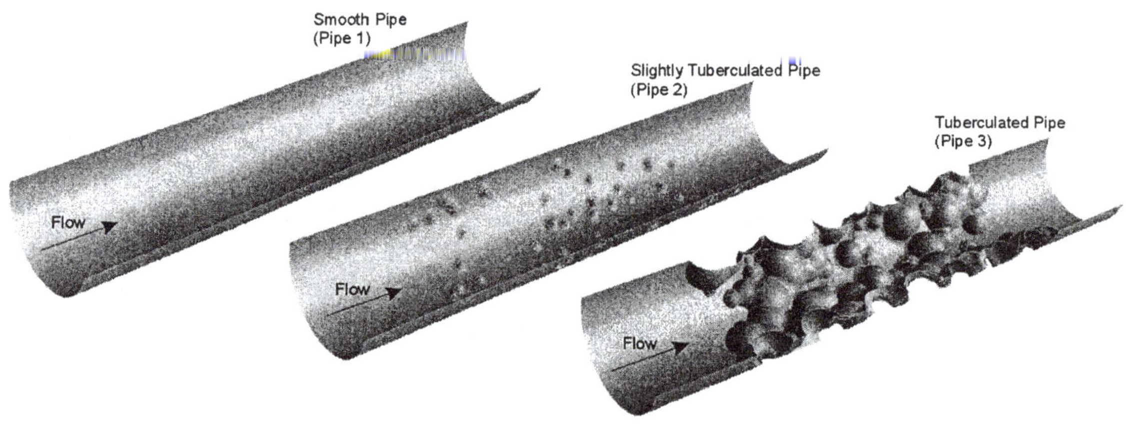

Figure 3.1 Cutaway pipes showing the three tuberculation states studied

RESULTS

CFD modeling allowed for the derivation of lifting velocities for particles of different size and specific gravity in mains of different diameter and states of tuberculation. The data are presented by showing the fractional pipe area cleaned as a function of bulk flow velocity, as illustrated in Figures 3.2 to 3.5. Flow velocities in the range 1.3–4.6 fps (0.4–1.4 m/s) were analyzed.

Figures 3.6 to 3.8 show a particular scenario of the performance of the 4-inch (100-mm) pipe. This example has been included to visually demonstrate how data in Figures 3.2–3.5 were derived. This scenario simulated the lifting of 120-µm diameter particles from the pipe wall with a mean velocity of 3.3 fps (1.0 m/s). In the dark areas, there is insufficient flow to lift the particles. These areas are generally downstream of the tubercles and in the recessed areas between tubercles. This type of output can be used to determine the fractional pipe area cleaned.

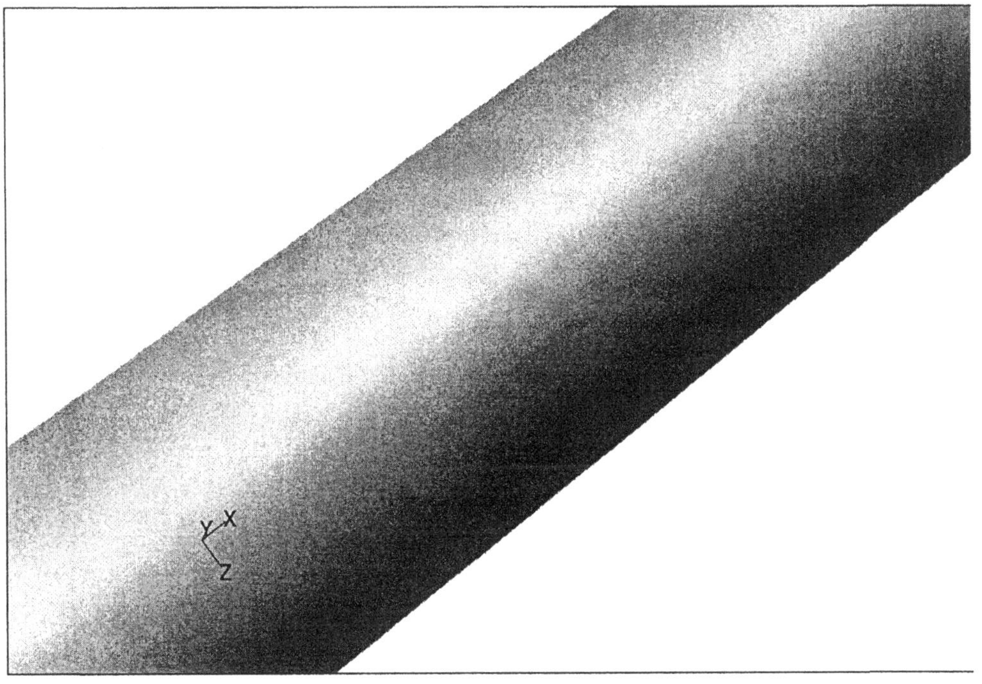

Figure 3.6 Smooth pipe showing uncleaned areas with shading

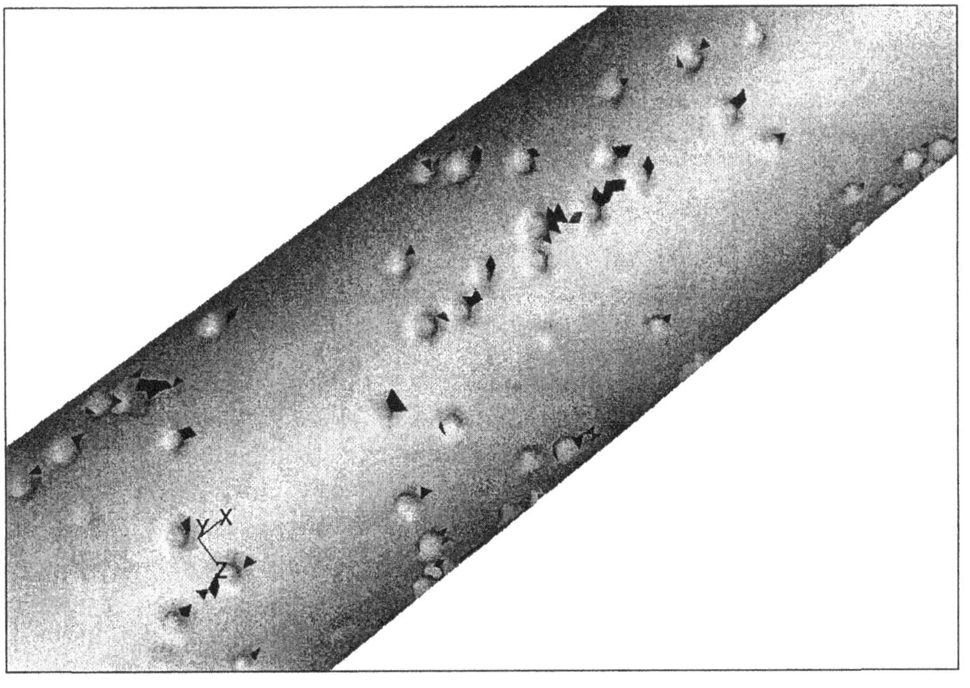

Figure 3.7 Slightly tuberculated pipe showing uncleaned areas with shading

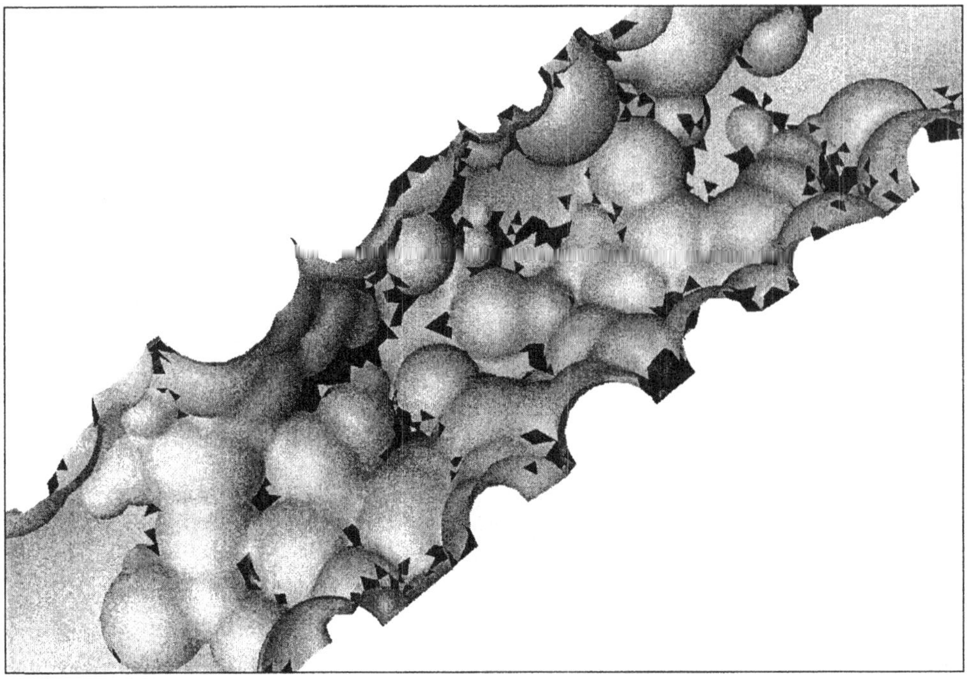

Figure 3.8 Tuberculated pipe showing uncleaned areas with shading

DISCUSSION

The complete CFD results have been incorporated into a flushing decision tree tool (see Chapter 7). Significant findings are discussed below.

Pipe Diameter

Pipe diameter might be expected to influence the lifting of particles. This is because, for a larger pipe, the increase in Reynolds number reflects greater turbulence. More turbulent flow will create higher velocities close to the pipe wall, with corresponding increases in velocity gradients, and hence increased lift forces. However, the results do not show a significant difference between the 4-inch (100-mm), 6-inch (150-mm), and 8-inch (200-mm) mains. This indicates that nominal pipe diameter does not significantly affect particle removal performance (for the pipe sizes studied). This can be explained as follows. Doubling the pipe diameter from 4-inch (100-mm) to 8-inch (200-mm) only doubles the Reynolds numbers, which are all above 40,000, and hence already fully turbulent. An increase in Reynolds number of at least an order of magnitude would be required to produce any significant change in the near wall conditions. Since nominal pipe diameter does not significantly affect particle removal, the results have been amalgamated in Figure 3.5. The remaining discussion will be based on the averaged results.

Pipe Tuberculation

In the smooth pipe, particles with a specific gravity between 1.1 and 3.0 were removed by flushing at mean velocities of 2.0–2.6 fps (0.6–0.8 m/s). The particle removal performance shows a critical velocity range, above or below which particles will be either completely removed or not removed at all.

The slightly tuberculated pipe shows a similar profile, with slightly decreased particle removal across the entire flow range. Although the difference in performance is small, it is most marked at the higher flows, where it becomes very difficult to achieve complete particle removal.

The severely tuberculated pipe shows increased particle removal at lower velocities, but significantly decreased particle removal at higher flow velocities. In the tuberculated pipe, there are cavities and recesses between tubercles that are very difficult to clean, even with elevated mean pipe velocities. Increasing the mean flow velocity produces a diminishing increase in particle removal, so that complete cleaning of the tuberculated pipe is unlikely. However, there are also regions on the tops of the tubercles where the flow is significantly increased compared with the smoother pipes. These areas are more easily cleaned than the smooth pipe at low flows.

These results show that increased tuberculation generally reduces the cleaning efficiency of the flush. More importantly, it always reduces the maximum particle removal performance that is achievable by flushing, as there will always be recesses where flow velocity is negligible. This means that as the degree of tuberculation increases, the effectiveness of a flush becomes less sensitive to flushing velocity.

Particle Specific Gravity

Particle specific gravity has a significant effect upon particle removal. For slightly tuberculated mains, particles with a specific gravity of 3.0 required a 2.0 fps (0.6 m/s) higher mean velocity than particles with a specific gravity of 1.5.

In all cases, the fractional particle removal decreases as particle specific gravity increases. For the tuberculated cases a slightly lower fractional particle removal occurs at all flow rates. For the smooth pipes, the effect of increased specific gravity is to raise the minimum velocity required for complete lifting.

Particle Size

The results show that larger particles are always harder to remove than smaller ones. However, the sensitivity of fractional particle removal to particle diameter is low. In most cases, the additional velocity required to remove 200 μm particles rather than 20 μm particles is between 0.33 fps (0.1 m/s) and 0.66 fps (0.2 m/s).

Comparison to Previous Flushing Velocity Guidance

The CFD results predict significantly better particle removal than those of Stephenson (1989). For particles of specific gravity of 1.5 and 200 μm size in a 4-inch (100-mm) main, Stephenson (1989) recommends a minimum flushing velocity of 3.0 fps (0.9 m/s) for particle removal. In contrast, CFD modeling illustrated that flow velocities below 1.3 fps (0.4 m/s) would be sufficient for complete removal of these particles from smooth mains. For particles of specific gravity of 3.0 and 200 μm size in a 4-inch (100-mm) main, Stephenson (1989) recommends a minimum flushing velocity of 5.9 fps (1.8 m/s) for particle removal. However, CFD modeling illustrated that flow velocities of 2.0 fps (0.6 m/s) were sufficient to scour these particles from smooth surfaces.

The models developed around the smooth pipe agree well with the 'rule-of-thumb' value provided by Williams (2000). This rule-of-thumb, which is based upon previous pipe rig data, suggests that a mean velocity of 1.6 fps (0.5 m/s) is required to suspend most loose sediment. However, the CFD data suggests it should not be used for calculating flush rates for tuberculated pipes.

CONCLUSIONS

CFD modelling has been applied to quantify the flushing velocities needed to lift loose particles of various sizes and densities. The findings of the CFD modeling show that the most significant factors in determining the flush rate are the extent of pipe tuberculation and particle density. Particle size is less significant (over the 20 to 200 μm range) and the pipe diameter has almost no effect. Specific findings from CFD modelling include:

(1) Pipe tuberculation has a significant effect upon flushing, reducing the achievable removal performance. Increasing pipe tuberculation makes flushing performance less sensitive to flushing velocity.

(2) Particle specific gravity has a significant effect upon removal.

(3) Nominal pipe diameter does not significantly affect particle removal performance for the pipes studied (4", 6" and 8").

(4) The removal of larger particles requires slightly higher flushing velocities: 200μm particles required up to 0.2m/s (0.66 ft/s) higher mean velocity, compared to 20μm particles.

(5) Current industry flush rates are higher than required.

CHAPTER 4
PILOT STUDY PROGRAM

INTRODUCTION

The aim of the pilot study program was to determine, in a controlled environment, the flushing velocities required for lifting and removal of different types of contaminants. A second objective was to compare the lifting velocities derived from experimental testing to those derived from CFD modeling. A flushing rig was built for the experimental testing. The findings are detailed in this chapter.

BACKGROUND

From the literature review, the following categorization system was adopted for different types of material in a pipeline environment:

- Loose particles recently settled or generated on the invert of the pipe;
- Cohesive deposits settled or generated on the invert of the pipe which form a compact, complex deposit;
- Adherent deposits generated on and attached to the invert of the pipe.

Contaminants representative of these different categories were selected for the pilot scale flushing trials. The contaminants listed in Table 4.1 were selected to represent the most common type of particle or deposit found in distribution systems for each category. GAC was included since it is now a common element in the treatment process and GAC fines have the potential to become a more prevalent type of distribution system deposit.

Table 4.1
Materials utilized in pilot studies

Loose deposits	Cohesive deposits	Adherent deposits
GAC particles	Iron corrosion sludge	Iron tubercles
Sand	Ferric floc	
Iron corrosion particles	Manganese coating	
	GAC clumps	

METHODOLOGY

Pipe Test Rig

The experimental pipeline (test rig) provides an experimental facility to test the effect of different flow rates on particle/deposit lifting. The experimental pipeline is based at Thames Water's Research & Technology Centre at Kempton Park Treatment works in SW London. The rig is constructed over ground to allow visual observations to be made. Test sections can be easily installed into the rig. Tapping points enable particles to be inserted, or seeded, into the rig.

A schematic of the pipe test rig is provided as Figure 4.1. The rig consists of 13-ft (4-m) of clear 4-inch (100-mm) diameter test pipe, with a hydrant at the downstream end. A pump recirculates the flow via a tank. Particles can be introduced into the pipe via a tapping at the upstream end of the pipe. By opening the hydrant, the rig can be flushed at flow rates up to approximately 350 gpm (22 L/s), corresponding to flow velocities up to 9.1 fps (2.8 m/s). This maximum flow rate represents an aggressive flush, and is considerably higher than current flushing flow rates achievable at many locations in the field.

Nets on the discharge hose and a fine mesh on the inlet pipe prevent particles from being recirculated through the rig. A flow meter immediately upstream of the test pipe records the flow rate. A turbidity meter records the turbidity of the water entering the test pipe. A second identical turbidity meters monitors the water 1 meter (3.28 feet) upstream of the hydrant. Both turbidity meters log measurements at 35-second intervals.

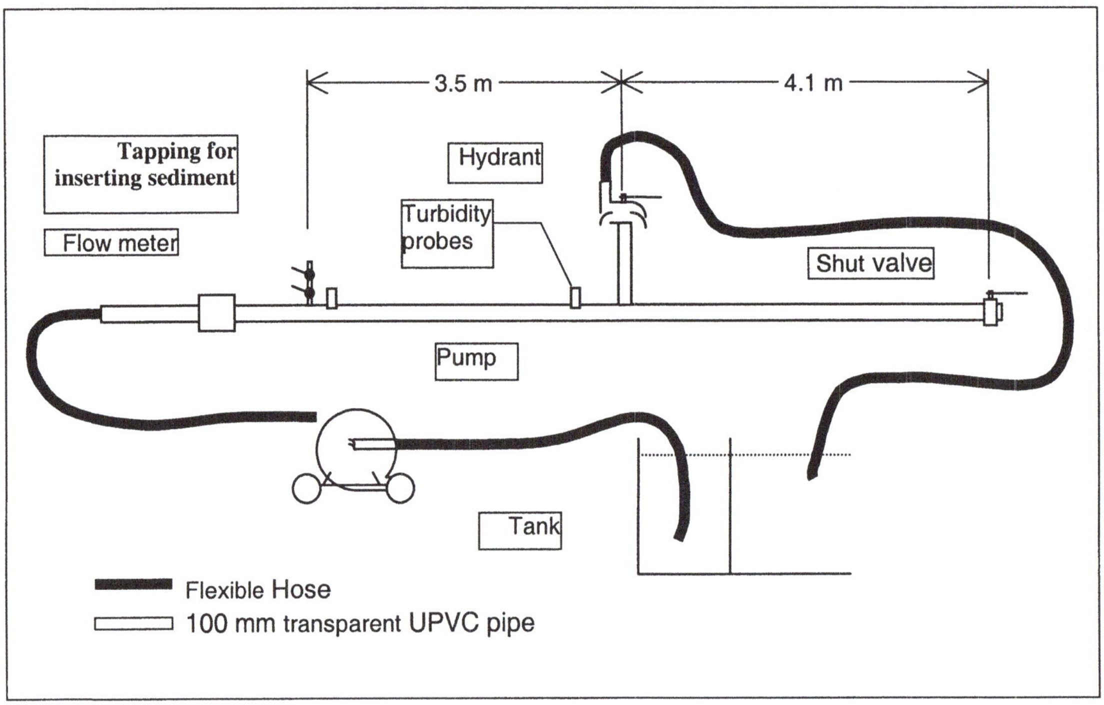

Figure 4.1 Pipe test rig schematic

A summary of the water quality supplied to the pipe test rig is given in Table 4.2.

Table 4.2
Summary of water quality supplied to pipe rig

Parameter	Range	Standard deviation	No. of samples
PH	7.68 - 8.08	0.11	11
Total calcium (mg/L)	104 – 117	3.78	11
Alkalinity (mg/L)	204 – 222	5.6	11
Chloride (mg/L)	44 – 49	1.77	11
Sulfate (mg/L)	61.1 - 66.7	1.49	11
Total chlorine (mg/L)	0.416 - 0.713	0.09	11
Free chlorine (mg/L)	0.024 - 0.101	0.028	11
Calcium carbonate precipitation potential (mg/L)	25.48 - 33.09	2.63	11

Loose Particle Flushing Trials

Examples of three different types of distribution particles were obtained: sand, GAC, and iron corrosion particles. The sand was obtained from a slow sand filter bed at a Thames Water treatment plant in South West London. The GAC was obtained from a slow sand filter bed incorporating a GAC sandwich. The iron corrosion deposits were collected during routine flushing operations in a network composed of predominantly unlined cast iron pipes. Particles were captured in 57-μm nets. Glass microspheres were also tested as a control.

Prior to insertion into the rig, all particles were analyzed for size and specific gravity. This provided sub-samples of particles of known size and specific gravity for testing. Once seeded, the particles were left to settle. Next, the flow was gradually increased from 0 fps by opening the hydrant, and the particles were observed for the following:

- First visual movement
- Shuffling movement of individual particles (along the bottom of the main)
- Shuffling movement of all of the particles (along the bottom of the main)
- Lifting of all particles into the main flow

For each of the movement types listed above, the velocity at which it occurred was recorded. The lifting experiments of each different particle size range and density, for each different particle type, were repeated at least three times. This provided mean velocities derived from at least three trials.

Cohesive Deposit Flushing Trials

Four different types of cohesive deposits were obtained: iron corrosion sludge, manganese, GAC clumps, and ferric floc. Experiments involved incorporating the cohesive deposits in the pipe test rig, incubating the deposit to facilitate cohesion to different pipe materials, operating different velocities, and then observing the velocities associated with particle lifting events.

Iron Corrosion Sludge

Test sections were obtained from a 0.8-mile (1.3-km) pipe rig (TORUS) owned and operated by Thames Water (Holt et al. 1994). The sections were composed of unlined cast iron and had been in place for at least a 2-year period, allowing the development of corrosion deposits and iron sludge on the pipe walls. Test sections were removed from two parts of the rig: one where velocity was very low (0.02 fps) but continuous, and another where flow was intermittent with periods of regular stagnation (known as the stagnation section). Iron sludge deposits had developed on all test sections. Most iron sludge was found on the test sections exposed to the intermittent flow regime.

The flushing methodology used for these trials is described in detail in Appendix C. In summary, the iron test pipes were inserted into the rig between clear PVC sections and flow was increased in 16 gpm (1 L/s) steps from 0 to 350 gpm (22 L/s). Each flow rate was maintained for approximately 3 to 5 minutes after which the hydrant net was examined and cleaned, and any captured particles were collected for analysis. The method for capturing the lifting event was visual (i.e., observation of the occurrence of discoloration) and by the capture of particles in downstream nets.

Manganese Coating

A section of PVC pipe coated in manganese was exhumed from a distribution system in the Netherlands. A summary of the water quality supplied to the pipe in the distribution system is given in Table 4.3. There was a history of manganese related complaints in the area from which the pipe was exhumed.

Table 4.3
Summary of water quality supplied to manganese-coated pipe in the Netherlands

Parameter	Minimum	Average	Maximum
PH	7.78	7.84	8.00
Total calcium (mg/L)	71.5	72.0	72.6
Alkalinity (mg/L)	253	257	266
Chloride (mg/L)	9.2	9.6	10.6
Sulfate (mg/L)	<2	<2	<2
Manganese (ug/L)	0.008	0.025	0.143

The pipe section was cut in half to enable a repeat of the trial. The trial methodology used was the same as that for the iron sludge trials and is described in detail in Appendix C. The main method for capturing the sloughing event was visual, including observation of black particles downstream of test sections, and the capture of particles in downstream nets. Chemical analysis of the black particles in the hydrant nets was also carried out to confirm the source of the particles.

GAC Clumps

GAC was obtained from a slow sand filter bed filter bed incorporating a GAC sandwich at a Thames Water treatment plant in South West London. GAC was seeded into the rig via a tapping point and left in contact with test sections for several days. It was anticipated that the incubation time would facilitate cohesion of the GAC granules with each other and with the pipe wall. Iron and PVC test sections were trialed separately. The trials involved gradually stepping up the flow until a lifting event was observed. The trial methodology for GAC fines is described in detail in Appendix C. The main method for capturing the lifting event was visual (i.e., observing the lifting of particles and the capture of particles in downstream nets). Chemical analysis of GAC particles in hydrant net was not warranted since GAC particles were distinctive and thus easy to identify in a hydrant net.

Ferric Floc

The ferric floc was made with ferric sulfate and the raw water feeding the treatment plant using standard laboratory jar test equipment. Floc was inserted into the rig using tapping points and was left for several days to facilitate cohesion prior to the trials. Ferric floc trials were performed using both iron and PVC as the test pipe sections. The trial methodology for ferric floc trials is described in detail in Appendix C. Sloughing events were captured by observation. Entrainment of iron flocs was easy to capture by eye. A summary of the different trials performed is given in Table 4.4. The diameter of all test pipe sections used was 4 inches (100 mm).

Table 4.4
Summary of cohesive deposit flushing trials

Deposit-trial	Test section material	Source of pipe and deposit	Incubation (days)
Iron sludge 1	Iron	Stagnation section of TORUS pipe rig	1
Iron sludge 2	Iron	TORUS (continuous flow section)	Several hours
Iron sludge 3	Iron	TORUS (continuous flow section)	1
GAC 1	Iron	TORUS (stagnation section) GAC from filters	6
GAC 2	Iron	TORUS (stagnation section) GAC from filters	5
GAC 3	Iron	TORUS (stagnation section) GAC from filters	5 minutes
GAC 4	PVC	Thames Water pipe GAC from filters	4
GAC 5	PVC	Thames Water pipe GAC from filters	2

(continued)

Table 4.4 (continued)

Ferric floc 1	Iron	TORUS (stagnation section) floc made in jar test	3
Ferric floc 2	Iron	TORUS (stagnation section) floc made in jar test	4
Ferric floc 3	PVC	Thames Water pipe Floc made in jar test	4
Manganese 1	PVC	(KIWA) Amsterdam distribution system	5
Manganese 2	PVC	(KIWA) Amsterdam distribution system	3
Manganese 3	PVC	(KIWA) Amsterdam distribution system	Several hours

Adherent Deposit Flushing Trials

Unlined cast iron pipe sections with iron tuberculation were used in the experiment. The pipe sections were all 4-inch (100-mm) in diameter, cast iron material, exhumed from Thames Water's distribution system. They were selected because they provided examples of typical iron corrosion deposits found in the distribution system. The pipe sections were installed between the flanges prior to commencing trials. Clear PVC pipe was installed upstream and downstream of the test section. Details of the methodology are given in Appendix C. In summary, the trials involved increasing the flow in the test section in 16 gpm (1 L/s) steps up to a flow rate of 350 gpm (22 L/s) and observing and capturing the particles sloughed off at each flow increment. The trials were repeated four times using different test sections each time.

Analysis

Particles were analyzed for particle size and specific gravity. These were considered to be the key parameters determining particle suspension and transport. The analysis was carried out at University of Surrey, Material Characterization Laboratory. Particle size was determined by wet sieving according to BS1377 (Part 2: 1990, Wet Sieving Method) using a range of different sieve sizes. Average specific gravity of the different types of sediment in each sieved size range was determined according to BS1377 (Part 2: 1990, Determination of Particle Density- small pyknometer method).

RESULTS

Loose Deposit Trials

GAC Particles

It was easy to observe the movement of the GAC particles due to their color (black) and because the particles were mostly greater than 500-μm in size. When all of the GAC particles were initially moving, they tended to shuffle along the bottom of the main as one large group. Results are summarized in Table 4.5 and Figure 4.2.

Table 4.5
Results from pipe test rig lifting velocity trials - GAC particles

Sieve size (micron)	Mass (g)	SG (Mg/m^3)	First visual movement (ft/sec)	Shuffling movement (ft/sec)	All particles moved – shuffle (ft/sec)	All particles moved - lifted (ft/sec)
<500	9.8	-	0.21 ± 0.00	0.23 ± 0.00	0.70 ± 0.00	1.48 ± 0.02
500	61.7	1.699	0.24 ± 0.00	0.28 ± 0.02	0.70 ± 0.00	1.51 ± 0.04
600	165.8	1.678	0.28 ± 0.00	0.37 ± 0.04	0.74 ± 0.00	1.52 ± 0.04
710	546.3	1.634	0.34 ± 0.00	0.43 ± 0.04	0.71 ± 0.00	1.59 ± 0.02
850	168.6	1.624	0.36 ± 0.00	0.46 ± 0.04	0.72 ± 0.00	1.62 ± 0.01
1000	112.1	1.608	0.38 ± 0.00	0.51 ± 0.02	0.79 ± 0.00	1.65 ± 0.01
1180	72.4	1.637	0.40 ± 0.00	0.56 ± 0.02	0.79 ± 0.00	1.66 ± 0.00
1400	22.2	1.638	0.43 ± 0.00	0.58 ± 0.04	0.90 ± 0.00	1.71 ± 0.04

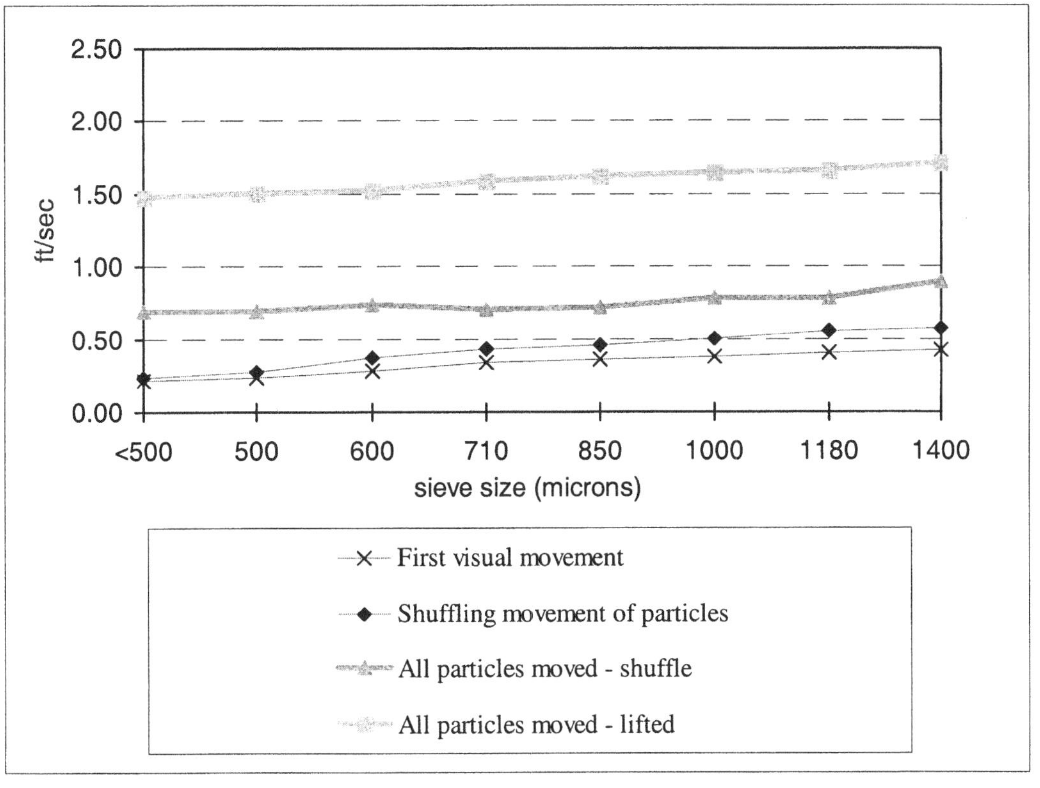

Figure 4.2 Summary of velocities required to move and lift GAC particles

Sand Particles

Sand was more difficult to observe than GAC since the sample was composed of smaller particles (less than 90-μm to 400-μm) and the color was less visible. Initial and shuffling movements were reasonably easy to record and the style of movement was similar to that of the GAC. Once moving, the sand tended to settle into a series of 'bars' that were orientated along the floor of the main normal to the direction of flow.

Table 4.6
Results from pipe test rig lifting velocity trials - sand particles

Sieve size (micron)	Mass (g)	SG (Mg/m^3)	First visual movement (ft/sec)	Shuffling movement (ft/sec)	All particles moved – shuffle (ft/sec)	All particles moved – lifted (ft/sec)
<90	80.3	2.746	0.42 ± 0.10	0.79 ± 0.25	1.06 ± 0.25	1.80 ± 0.18
90	210	2.699	0.47 ± 0.04	0.76 ± 0.20	0.91 ± 0.11	1.80 ± 0.13
125	55.7	2.732	0.42 ± 0.03	0.76 ± 0.28	0.86 ± 0.10	1.84 ± 0.10
150	218.2	2.755	0.45 ± 0.05	0.74 ± 0.25	0.87 ± 0.05	1.86 ± 0.11
180	242.1	2.708	0.49 ± 0.07	0.80 ± 0.18	0.95 ± 0.19	1.88 ± 0.10
212	83.4	2.714	0.42 ± 0.09	0.74 ± 0.11	0.91 ± 0.07	1.87 ± 0.14
250	86.9	2.745	0.55 ± 0.02	0.81 ± 0.16	0.91 ± 0.16	1.91 ± 0.13
300	33.5	2.743	-	-	-	-
355	5.6	-	0.47 ± 0.06	0.75 ± 0.09	0.93 ± 0.23	1.83 ± 0.21
400	-	-	0.43 ± 0.01	0.67 ± 0.11	0.82 ± 0.10	1.70 ± 0.15

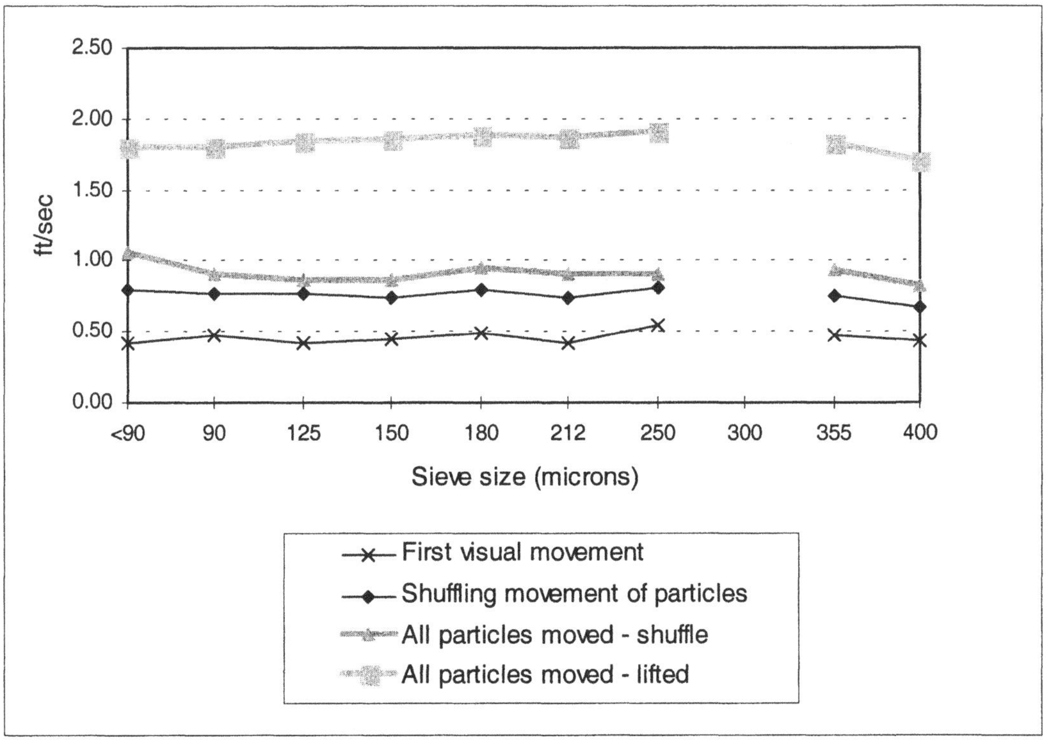

Figure 4.3 Summary of velocities required to move and lift sand particles

Iron Corrosion Particles

The iron corrosion particles contained a subset of fine particles (<45 μm). First visual movement of individual fine particles could not be accurately observed. However, first visual movement of the mass of fine particles could be observed and the associated velocity accurately quantified. For the particle sizes ranging from <45 μm to 1,180 μm, the movement style of the particles was similar to that of the sand. A subset of large particles (i.e., pebbles and stones >2,360 μm) was present in the sample. These particles were not tested in this series of experiments.

Table 4.7
Results from the pipe rig trials - distribution system particles

Sieve size (micron)	Mass (g)	SG (Mg/m^3)	First visual movement (ft/sec)	Shuffling movement (ft/sec)	All particles moved – shuffle (ft/sec)	All particles moved – lifted (ft/sec)
<45	17.8	3.112	0.44 ± 0.01	0.44 ± 0.01	0.45 ± 0.00	1.58 ± 0.00
45	37.7	2.808	0.34 ± 0.00	0.44 ± 0.04	0.49 ± 0.00	1.64 ± 0.00
75	103.4	2.935	0.23 ± 0.01	0.27 ± 0.02	0.48 ± 0.01	1.63 ± 0.01
150	121.3	2.706	0.19 ± 0.01	0.23 ± 0.02	0.52 ± 0.01	1.78 ± 0.11
300	114.7	2.649	0.10 ± 0.01	0.12 ± 0.00	0.51 ± 0.01	1.93 ± 0.03
600	89.3	2.654	0.07 ± 0.01	0.12 ± 0.00	0.69 ± 0.03	2.17 ± 0.11
1180	77.1	2.944	0.08 ± 0.01	0.30 ± 0.01	1.10 ± 0.02	2.36 ± 0.06

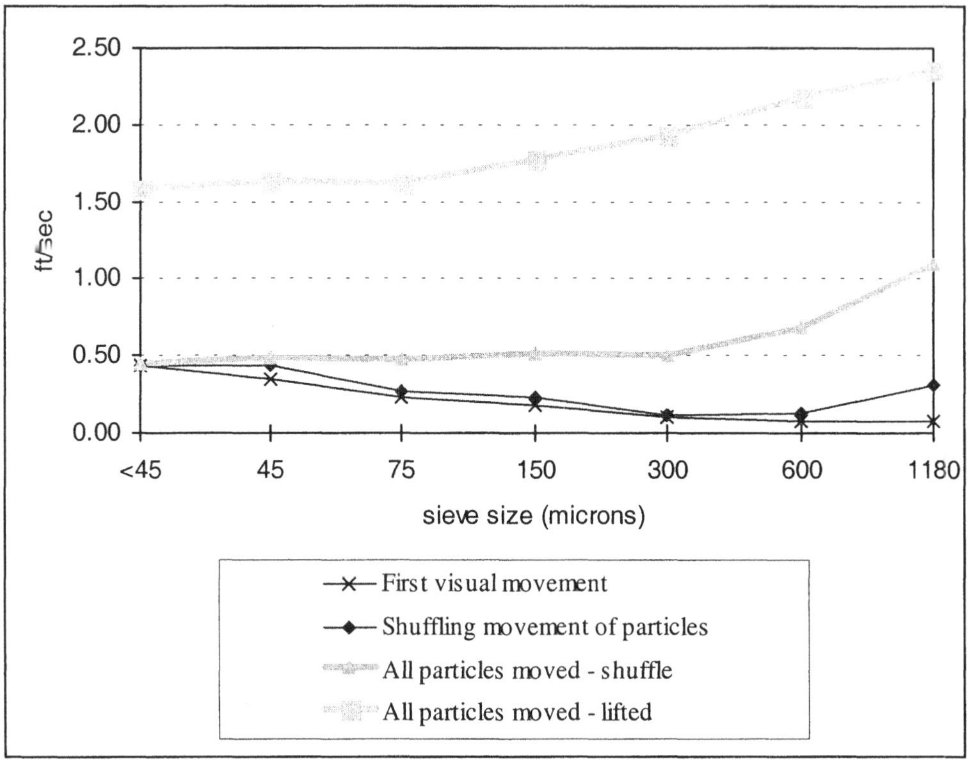

Figure 4.4 Summary of velocities required to move and lift distribution system particles

Glass Microspheres

Glass microspheres were tested to provide additional data on particle size and density versus lifting velocity.

Table 4.8
Results from pipe test rig lifting velocity trials - glass microspheres

Sieve size (micron)	Mass (g)	SG (Mg/m^3)	First visual movement (ft/sec)	Shuffling movement (ft/sec)	All particles moved – shuffle (ft/sec)	All particles moved – lifted (ft/sec)
250	-	2.2 - 2.65	0.13	0.18	0.58	2.14

Cohesive Deposit Trials

Iron Corrosion Sludge

The behavior of iron sludge during flushing was best demonstrated by the following:

- Observations on the aesthetic deterioration of water quality downstream of test pipe sections;
- Photographs of the pipe sections before and after the trials were performed; and
- Particles collected in the hydrant net.

Observations downstream. The iron sludge was removed from test pipe 1 (from stagnation section) as soon as the pump was turned on to start the trial, thus at very low velocities (<0.43 fps [0.1 m/s]). Orange discoloration of the water downstream of the test pipe was observed at this time. No subsequent discoloration, however, was observed downstream of the pipe during the trial. No discoloration was seen downstream from test pipes 2 and 3 (from continuous flow regime) throughout the trials.

Photographic evidence. Figures 4.5 and 4.6 show photographs of iron test pipe 1 before and after it was flushed, respectively. As can be seen in the photographs, there was no soft layer of sludge remaining on the pipe after the flushing was performed. Only the hard tuberculation layer remained. Figures 4.7 and 4.8 show iron test pipe 2 before and after its flushing trial was performed, respectively. Again, only the adherent iron tuberculation was left at the end of the trial. The third iron test pipe was also similar to iron pipe 2 before and after it was flushed.

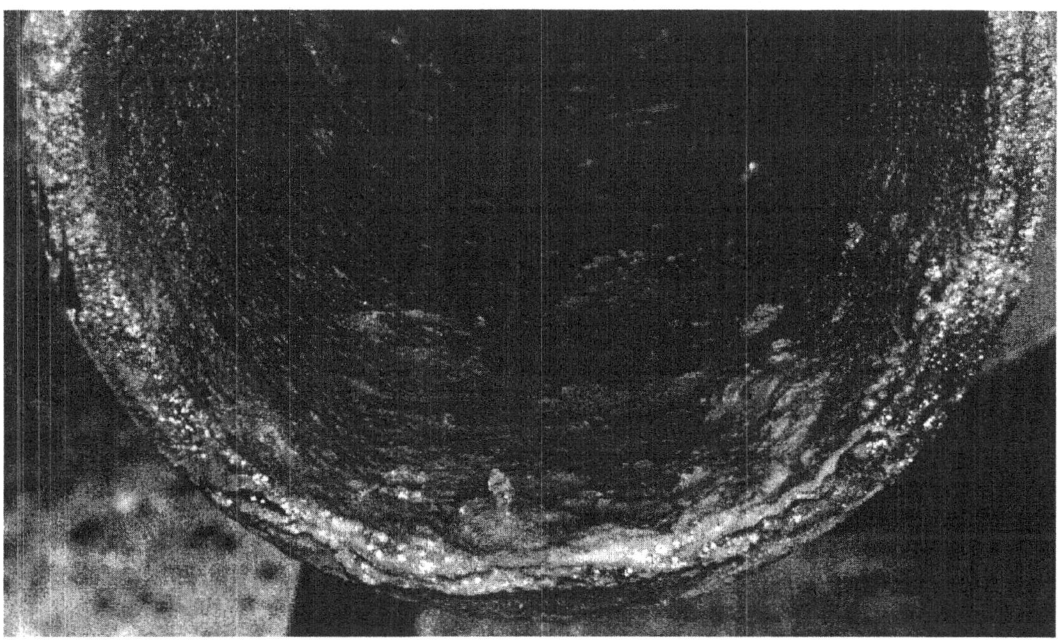

Figure 4.5 Iron test section 1 before flushing trial performed

Figure 4.6 Iron test section 1 after flushing trial performed

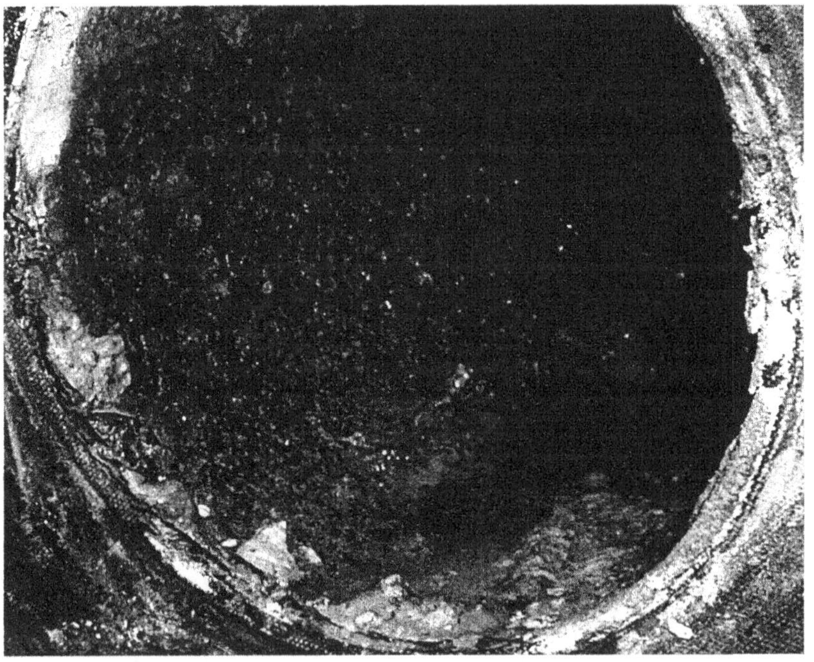

Figure 4.7 Iron test section 2 before flushing trial performed

Table 4.10
Summary of results from GAC flushing trials

Pipe / Trial	1	2	3	4	5
Test pipe section material	Iron	Iron	Iron	PVC	PVC
Time in rig before trial	6 days	5 days	5 min	5 days	2 days
Lifting flush rate (L/s)	4.7	2.8	2.5	9.2	4.2
Lifting flush rate (gpm)	74.8	44	39.6	145.3	66
Lifting velocity (m/s)	0.6	0.4	0.3	1.2	0.5
Lifting velocity (fps)	1.968	1.31	0.98	3.946	1.64

Photographic evidence. Figure 4.12 shows GAC moving up the hydrant during a trial (GAC trial 5: 66 gpm [4.2 L/s]) after it had been lifted from the test pipe.

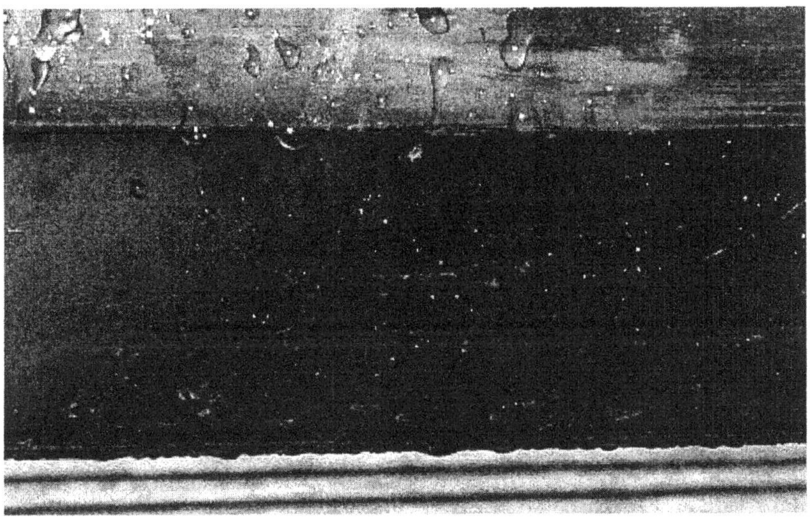

Figure 4.12 Example of GAC lifting up the hydrant

Particles collected in the hydrant net. Particles were collected in the nets only at certain flow rates. The amount of GAC captured in the hydrant nets at these flow rates was measured (wet weight) and was roughly equal to the amounts of GAC inserted into the rig through the tapping points before the trials. This indicates that all the GAC was removed at set flush rates.

The following are key findings from the GAC clump flushing trials.

- GAC inserted into the rig was effectively removed by flushing.
- The sloughing of GAC occurred at discrete flushing velocities.
- The five trials provided evidence to indicate that GAC can form a complex deposit in different pipeline environments, exposed to different periods of stagnation:
  - For iron pipes, the velocities required to lift deposits doubled after the period of incubation was increased from 5 minutes to 6 days. However, all deposits were lifted at a velocity of 2.0 fps (0.6 m/s). This velocity agrees with CFD modeling

results, which estimate that most loose material of similar density to GAC (specific gravity 1.65) in slightly tuberculated pipes will be removed by 2 fps (0.6 m/s) flush. The comparison of observed to CFD-derived lifting velocity suggests that GAC formed a loose deposit in the iron pipes tested.

- For PVC pipes, after 5 days incubation, the GAC lifted at approximately double the velocity predicted for loose deposits of similar density in smooth pipes. This observation suggests that GAC particles require higher velocities to remove than that predicted by modeling of loose particles. The comparison of observed to CFD-derived lifting velocity suggests that GAC formed a complex, cohesive deposit in PVC pipes after 5 days incubation/stagnation.

Ferric Floc

The behavior of ferric floc on pipe walls during flushing was best demonstrated by observations and photographs of discoloration events downstream of the test pipe sections due to floc entrainment.

Observations downstream. In each of the ferric floc trials performed, the lifting event was distinct and dramatic, and discoloration was observed downstream of the test pipe. The ferric floc was always seen to suddenly lift and move downstream of the test pipe and up the hydrant. The flushing velocities at which these events were observed are summarized in Table 4.11.

Table 4.11
Summary of results from ferric floc flushing trials

Pipe / Trial	1	2	3
Test pipe section material	Iron	Iron	PVC
Time in rig before trial	3 days	4 days	4 days
Lifting flush rate (L/s)	1.7	2.2	2.2
Lifting flush rate (gpm)	26.4	35.2	35.2
Lifting velocity (m/s)	0.21	0.28	0.28
Lifting velocity (fps)	0.69	0.92	0.92

Photographic evidence. Figure 4.13 shows an example of the discoloration due to lifting and transportation of ferric floc during the trials. Figure 4.14 shows the same pipe with clear water flowing through it for comparison.

Figure 4.15 Photograph of example test section before stepped flushing performed

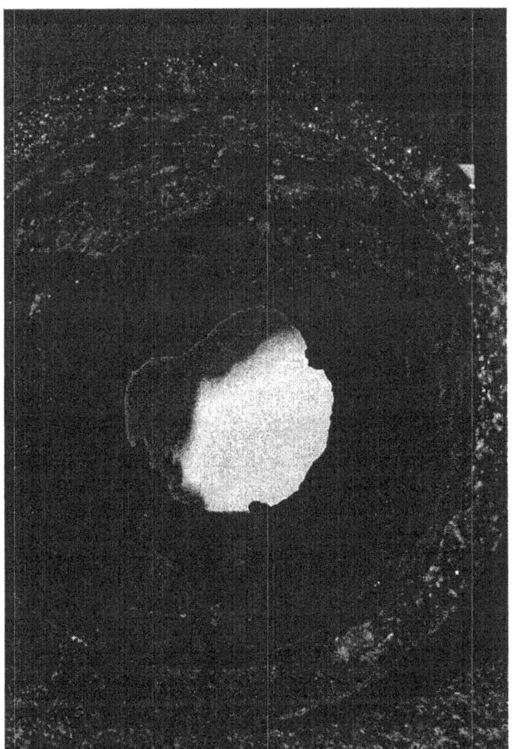

Figure 4.16 Photograph of example test section after stepped flushing performed

Particles collected in the hydrant net. The occurrence of particles in hydrant nets downstream of the test section after each step increase in flow was recorded. Although particles were sometimes present in the hydrant nets after each increase in flush rate, the number of particles present was very low. The quantities were sufficiently low to prevent weighing. No visible discoloration was observed associated with the presence of particles. There was also no consistent pattern of presence/absence of particles associated with step increases in flow.

In summary, the trials with iron tubercles provided no evidence of a significant sloughing event associated with hydrant flushing. Therefore, flushing, even at flow rates higher than those achievable in the field, did not produce the shear velocities required to dislodge adherent iron deposits from the test sections trialed.

CONCLUSIONS

Table 4.12 summarizes the lifting velocities recorded in the pilot scale trials for the different types of contaminants.

Table 4.12

Summary of lifting velocities for different contaminants from experimental flushing trials

Contaminant category	Contaminant	Range of lifting velocities recorded during different trials (fps)
Loose deposits	Sand	1.70 - 1.91
	GAC	1.48 - 1.71
	Iron corrosion	1.58 - 2.36
Cohesive deposits	Iron corrosion sludge	< 0.43
	Manganese coating	No lifting observed
	GAC clumps	0.98 - 3.94
	Ferric floc	0.69 - 0.92
Adherent deposits	Iron tubercles	No lifting observed

The trials have demonstrated that different types of contaminants have different degrees of adherence and cohesiveness, impacting on the effectiveness of flushing as a control measure. Flushing may not be effective for removing manganese coatings or iron tubercles. Most of the contaminants tested would be removed by adopting flush rates determined by CFD modeling. Specific conclusions are cited below.

- **Sand, GAC, Iron corrosion particles.** These particles are commonly found in distribution systems and are amenable to removal by low level flushing at rates determined by CFD modeling.

- **Iron corrosion sludge.** This type of deposit can develop on unlined cast iron pipes. Iron sludge forms a very loose deposit and can be easily lifted, therefore amenable to removal by low level flushing. The flow rates for removal of loose deposits predicted by CFD modeling will remove iron sludge. The results indicate that iron sludge can accumulate in sufficient quantities to cause discoloration in parts of the network with unlined iron pipes where periodic stagnation occurs. In this scenario, there is a potential for discoloration and thus preventive flushing may be considered prior to any network activities involving hydraulic changes.

- **Manganese coating.** This type of deposit formed a very adherent coating on PVC pipe, not amenable to removal by flushing. Alternative, more aggressive control measures are required to remove this type of contaminant (e.g., swabbing, rehabilitation).

- **GAC fines**. Trials indicate that GAC can form a complex deposit in pipe networks. GAC deposits can exhibit loose or cohesive properties, influenced by factors such as material type and length of stagnation periods. The trials showed that for iron pipes, the GAC formed a loose deposit amenable to removal by flushing at flow rates predicted by CFD modeling. For PVC pipes, GAC formed a complex deposit on pipe walls and required a higher velocity to remove.

- **Ferric floc.** Ferric floc forms a very loose deposit and can be easily lifted. Therefore it is amenable to removal by low level flushing. The flow rates for removal of loose deposits predicted by CFD modeling will remove ferric floc. Ferric floc is likely to cause discoloration events since it can be easily entrained and must be removed prior to any work involving hydraulic changes to the network.

- **Iron tubercles.** This type of deposit formed a very adherent coating not amenable to removal by flushing. Alternative, more aggressive, control measures are required to remove this type of contaminant (e.g., swabbing, rehabilitation).

CHAPTER 5
FIELD TRIALS

INTRODUCTION

The third element of the project approach involved conducting distribution system flushing trials. The purpose of the field trials was to validate findings from CFD and pilot scale efforts, and to assess the effectiveness of flushing velocities in systems with varying pipe materials and source water qualities. This Chapter describes the field protocol and data analysis techniques used to determine:

- If more material was removed over a range of velocities,
- If different material was removed over a range of velocities, and
- What variables influenced or controlled the effectiveness of various flushing velocities.

Surveys were sent out to several utilities to gather information on drinking water quality, distribution system characteristics, historical water quality problems, and routine mains cleaning practices. Based on the responses received, a total of twelve water systems were selected to participate in the field studies. System locations and associated pipe types and water quality problems to be addressed through unidirectional flushing are summarized in Table 5.1. Finished water quality conditions are summarized for each utility in Table 5.2.

Table 5.1
Water systems participating in field trials

Utility location	Pipe type(s)	Water quality problems
Anchorage, AK (U.S.)	AC, UCI	Sediment, iron
Bellevue, WA (U.S.)	AC, UCI, CLDI	Sediment, discoloration
Cincinnati, OH (U.S.)	UCI	Red water
Cleveland, OH (U.S.)	CLCI	Phosphate, HPC
Dover, NH (U.S.)	UCI	Black water, HPC
Irvine Ranch, CA (U.S.)	PVC	Sediment
Newport News, VA (U.S.)	UCI, DI	Discoloration
Portland, OR (U.S.)	UCI	Discoloration
South Berwick, ME (U.S.)	UCI	Iron, manganese
Southern California (U.S.)	UCI	Taste & odor
Thames Water (U.K.)	PVC, UCI	Sediment
Yarra Valley (Australia)	UCI	Discoloration

AC=Asbestos Cement; UCI=Unlined Cast Iron; CLDI=Cement Lined Ductile Iron; CLCI=Cement Lined Cast Iron; PVC=Polyvinyl Chloride

Table 5.2
Finished water quality of participating utilities

Utility location	PH	Alkalinity (mg/L as $CaCO_3$)	Hardness (mg/L as $CaCO_3$)	Turbidity (NTU)
Anchorage, AK (U.S.)	7.2 – 8.8	54 – 144	68 – 192	0.20
Bellevue, WA (U.S.)	7.3 – 8.6	13 – 15	12 – 22	0.30
Cincinnati, OH (U.S.)	NA	NA	NA	NA
Cleveland, OH (U.S.)	7.2 – 7.5	78	114	NA
Dover, NH (U.S.)	7.2 – 7.4	14 – 33	15 – 40	NA
Irvine Ranch, CA (U.S.)	8.0	110 – 131	170 – 289	0.05 – 2.60
Newport News, VA (U.S.)	7.6	36	80	NA
Portland, OR (U.S.)	7.1 – 8.1	6 – 17	6	NA
South Berwick, ME (U.S.)	7.2 – 8.0	60 – 80	38	0.5
Southern California (U.S.)	8.0 – 8.3	93 – 131	142 – 289	0.04 – 0.08
Thames Water (U.K.)	7.7 – 8.1	204 – 222	260 – 353	0.1 – 2.0
Yarra Valley (Australia)	7.5 – 7.9	12 – 13	18 – 21	1.0 – 1.1

Utility location	Iron (mg/L)	Manganese (mg/L)	Calcium (mg/L)	Magnesium (mg/L)	Aluminum (mg/L)	Phosphate (mg/L)
Anchorage, AK (U.S.)	< 0.02	< 0.02	18 – 38	ND	ND	NA
Bellevue, WA (U.S.)	0.04 – 0.14	0.01	11 – 18	NA	0.03 – 0.06	NA
Cincinnati, OH (U.S.)	NA	NA	NA	NA	NA	NA
Cleveland, OH (U.S.)	0.04	ND	33	9	0.12	1.0
Dover, NH (U.S.)	0.05	ND – 0.15	5 – 16	< 0.2	< 0.1	0.8 – 1.6
Irvine Ranch, CA (U.S.)	NA	NA	57 – 73	22 – 27	0.11 – 0.26	NA
Newport News, VA (U.S.)	0.04	< 0.01	NA	NA	NA	0.23
Portland, OR (U.S.)	0.01 – 0.10	< 0.01	1 – 2	0.6 – 0.7	NA	NA
South Berwick, ME (U.S.)	0.27	0.19	NA	4	NA	NA
Southern California (U.S.)	NA	NA	33 - 73	15 – 27	0.05 – 0.26	NA
Thames Water (U.K.)	0.012-0.21	<0.003	104 – 117	NA	NA	NA
Yarra Valley (Australia)	0.08 – 0.10	< 0.01	4.7	NA	0.02 – 0.03	NA

ND = Not Detected
NA = Data Not Available

An example unidirectional flushing protocol is provided in Appendix D. Detailed statistical analyses using box and whisker plots are included in Appendix E. Field trial protocols and key findings are discussed below.

FLUSHING AND SAMPLING PROTOCOL

The Project Team generated unidirectional flushing protocols based on system maps provided by each utility. Depending on the size of the distribution system, between one and three loops were selected for flushing. Each loop was further divided into several manageable steps, with each step requiring specific valve sequencing in order to direct water from a single clean source through a particular stretch of main.

The majority of the steps involved unidirectional flushing at a specific velocity and performing on-site analyses, including time-to-clear, turbidity, pH, and disinfectant residual measurements. These data were used to assess general water quality changes throughout the duration of a flushing event. In order to draw velocity-specific conclusions, sequential low and high velocity flushing trials were performed on select step(s) at each water system. The low and high velocity target ranges were 2-4.5 fps (0.6-1.4 m/s) and greater than 5 fps (1.5 m/s), respectively. Additional sampling conducted during these steps involved bulk water collection from the hydrant sample tap and particle capture from the discharge hose using hydrant nets. Pipe geometry data needed for data analysis were collected for each step. Flushing velocity was measured with a pitot gauge. Water clarity (i.e., turbidity) and disposal constraints were used to determine when to terminate a flushing event.

BULK WATER TRIALS

Bulk water quality samples were collected as part of the field tests to provide characterization and quantification of contaminants removed during unidirectional flushing. In total, 19 multi-velocity tests were conducted within 10 different water systems. Samples were collected in laboratory-prepared containers and analyzed by various local commercial laboratories. The broader results of bulk water analyses were used to help answer the question: "Does a higher flushing velocity remove more material?"

Summary of Testing

The process of selecting utilities and specific portions of the distribution system for these tests was guided by the intention of capturing a diverse range of water quality conditions, pipe materials, and flushing objectives.

Table 5.3 provides a comprehensive summary of bulk water trials and associated analyses. Each step involved sequential low and high velocity flushing events, and separate analyses were conducted on samples collected from each velocity.

Table 5.3
Testing summary for bulk water trials

Site	Step	Pipe type	Iron	Manganese	Calcium	Total suspended solids	Heterotrophic plate count	Orthophosphate	Total coliform
ANC	5	AC	×	×		×			
ANC	13	AC	×	×		×			
YVW	5	UCI	×				×		×
YVW	13	UCI	×				×		×
YVR	4	UCI	×						
YVR	10	UCI	×						
NNI	4	DI	×	×			×		
NNH	3	UCI	×	×			×		
BEL	3	AC	×	×	×		×		
POR	2	UCI	×	×			×		
POR	6	UCI	×	×			×		
POR	12	UCI	×	×			×		
SB	2	UCI	×	×			×		
SB	5	UCI	×	×			×		
DOV	3	UCI	×	×			×		
SCAL	7	CLCI	×				×		×
IR	3	PVC	×						
IR	8	PVC	×						
CLE	5	CLCI	×				×	×	

Bulk water samples were collected at the beginning, midpoint, and end of each velocity tested. The samples were collected either at the point of diffusion or from a sample tap installed on a spare port. The initial sample was collected after the feeder barrel had been cleared (10 seconds). The flush midpoint was estimated based on the results of on-site turbidity measurements and/or water appearance. The final sample was collected just prior to terminating the flush.

Data Validation

The conditions of each study were carefully evaluated to determine the appropriateness of using the results to address the issues of particle characterization and velocity effectiveness. Several data validation criteria were established in order to eliminate compromised data from further analysis. The bulk water tests were required to satisfy each of the conditions listed in Table 5.4 in order to be included in subsequent data analysis.

Table 5.4
Data validation criteria for bulk water trials

1. Flushing must be unidirectional and in accordance with the field protocol.
2. A minimum of one pipe volume must be displaced at each velocity.
3. The main must be flushed to clear (turbidity of less than 5 NTU) at each velocity.
4. The low velocity flush must reside in the range 2.0 – 4.5 fps (0.6 – 1.4 m/s).
5. The high velocity flush must reside in the range 5.0 – 9.0 fps (1.5 – 2.7 m/s).
6. The low and high velocity tests must differ by at least 1.5 fps (0.5 m/s).
7. The entire length of main must be of the same nominal diameter, excluding the effects of diameter reduction due to accumulated materials.

This validation process was performed to ensure data quality and pending conclusions. As a result of imposing these criteria, the size of the valid bulk water data set was reduced by 26%. Fourteen field trials met all seven criteria.

Statistical Data Analysis on Bulk Water

The original bulk water results were subjected to descriptive statistical testing. The intent of this exercise was to aggregate bulk water quality data from all field studies and ascertain the existence of global trends. The results are summarized for parameters measured on-site (Table 5.5) and by a commercial laboratory (Table 5.6).

Table 5.5
Descriptive statistics of on-site results from sequential low and high velocity flushing trials

	Time to clear (min)		Turbidity (NTU)	
	Low velocity	High velocity	Low velocity	High velocity
Mean	19.1	16.6	147.9	37.8
Standard error	5.3	4.1	79.6	21
Median	11.0	10.5	31.7	9.9
Standard deviation	19.8	15.4	297.7	78.7
Minimum	0.0	4.0	0.2	0.2
Maximum	64.0	51.0	1000.0	295
Count	14	14	14	14

Table 5.6
Descriptive statistics of laboratory results from sequential low and high velocity flushing trials

	Iron (mg/L)		Manganese (mg/L)		HPC (No/mL)	
	Low velocity	High velocity	Low velocity	High velocity	Low velocity	High velocity
Mean	14.2	2.0	0.5	0.1	2579.4	833.5
Standard error	8.0	0.8	0.2	0.0	1936.6	701.5
Median	2.4	0.9	0.3	0.1	410.0	40.5
Standard deviation	29.8	3.2	0.6	0.1	5477.4	1984.2
Minimum	0.0	0.0	0.0	0.1	5.0	3.0
Maximum	110.0	12.3	1.6	0.4	16000.0	5700.0
Count	14	14	7	7	8	8

Due to the large standard error associated with the small sample size, mean values of time to clear are not significantly different between the two velocity targets. Comparatively, the mean values of bulk water turbidity, iron, manganese, and HPC are consistently higher at the low flushing velocity relative to the ensuing high velocity. However, this does not necessarily imply that the high velocity flush failed to provide significant additional benefit.

Box-and-whisker plots (box-plots) included in Appendix E were developed to help determine the impact of source water quality, flushing velocity and pipe material on discharge water quality. Box-and-whisker plotting is a powerful method of visualizing data and identifying central tendencies. From each valid bulk water flush, maximum values of turbidity, iron, manganese and HPC bacteria were used to create the box-plots. The impacts of both source water quality and pipe material were evaluated. The box plots used to evaluate source water quality and flushing velocity indicate that the median values of each parameter (turbidity, iron, manganese, HPC) are not significantly different (within a 95% confidence region around the median) between the low and high velocities. The box plots used to evaluate pipe material show that for each of the bulk water parameters, the median values for DI, UCI, and AC pipe were not significantly different (based on an approximation of a 95% confidence region about the median). However, the median turbidity and iron levels collected from the field study with PVC piping were significantly less than those from the other pipes, particularly UCI. This result suggests that a significant source of iron particles may be unlined cast iron pipe (versus finished water contribution), as there was no UCI in the water system with PVC piping.

Data Manipulation

Raw bulk water data has limited use for making comparisons between and within utilities or for drawing specific conclusions about the removal efficiency of different velocities. This is due to the fact that the flow rates and elapsed flushing times were unique for each trial. Several data manipulation steps were performed to generate results that could be reliably compared. These steps are described below.

Converting Concentration to Mass

The first manipulation step involved converting the concentration results for each parameter to mass results. This conversion was accomplished by integrating raw concentration data over the duration of the flush, as shown in equation 5-1.

$$M_{i,j} = Q\int_{T_o}^{T_f} C_i \, dT \qquad (5\text{-}1)$$

Where:
$M_{i,j}$ = Total mass of parameter i at velocity j (mg)
C_i = Concentration of parameter i (mg/L)
Q = Flow rate (L/min)
T = Time (min)

Numerical integration was applied because the data sets were small and discrete. The Trapezoid Rule was favored over Simpson's Rule since the time interval between samples was variable. Figure 5.1 provides an illustration of the calculation method. The shaded area corresponds to the Trapezoid Rule numerical approximation of the integral.

$$M_{i,j} = Q\cdot\left[\left(\frac{C1+C2}{2}\right)(T2-T1)+\left(\frac{C2+C3}{2}\right)(T3-T2)\right] \qquad (5\text{-}2)$$

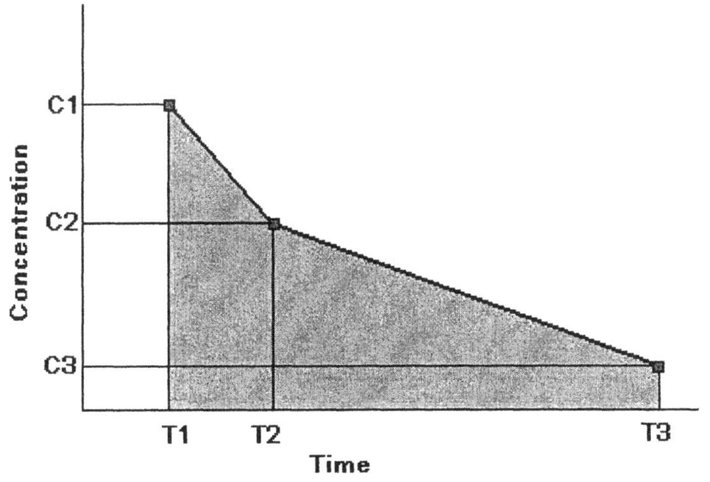

Figure 5.1 Trapezoid rule numerical integration

In some instances, the final sample was collected several pipe volumes after the water main had been flushed to clear. In these cases, the time when the last sample was collected (T3) was assigned to the elapsed time when the turbidity reached 5 NTU (based on linear interpolation). This manipulation was performed to minimize error associated with the numerical approximation.

Normalizing Mass to Surface Area of Main

The second data manipulation step involved normalizing the mass results to the total internal surface area of the main being flushed. The internal surface area was calculated assuming a smooth pipe surface with negligible diameter reduction due to accumulated deposits. This conversion was accomplished through the use of equation 5-3.

$$\hat{M}_{i,j} = \frac{12 \cdot M_{i,j}}{\pi DL} \qquad (5\text{-}3)$$

Where:

$\hat{M}_{i,j}$ = Normalized mass of parameter i at velocity j (mg/ft^2)

$M_{i,j}$ = Total mass of parameter i at velocity j (mg)

D = Diameter of main (inches)

L = Length of main (ft)

Development of Theoretical High

Since the low and high velocity tests were performed sequentially on the exact same stretch of main, any mass removed during the initial low velocity flush is not reflected in the high velocity mass result. However, if only the high velocity flush had been performed, the total mass removed should, in theory, approximate the sum of the low and high velocity mass results. The asterisk superscript (*) is used hereafter to refer to theoretical high velocity results.

$$M_i^* = M_{i,low} + M_{i,high} \qquad (5\text{-}4)$$

$$\hat{M}_i^* = \hat{M}_{i,low} + \hat{M}_{i,high} \qquad (5\text{-}5)$$

Where:

M_i^* = Theoretical mass of parameter i collected at high velocity (mg)

$\hat{M}_i^*$ = Normalized theoretical mass of parameter i collected at high velocity (mg/ft^2)

Low Velocity Mass Removal Efficiency

The intent of bulk water sampling was not to determine whether or not flushing could remove all of the loose and adherent deposits in distribution mains, but rather to identify how much material could be removed at the low velocity relative to the total amount that could be removed by high-velocity unidirectional flushing. The low velocity mass removal efficiency was calculated from equation 5-6.

$$E_i = \frac{M_{i,\,low}}{M_i^*} \times 100 \quad (5\text{-}6)$$

Where:
E_i = Low velocity mass removal efficiency for parameter i (%)

The low velocity removal efficiency is inversely related to the benefit of flushing at the high velocity. However, in order to capture the true benefit, the magnitude of the mass of material removed should be considered as well. Consider the following hypothetical example:

> In study Y, 0.5 grams of material are removed at the low velocity flush, and an additional 1.5 grams are removed during the ensuing high velocity flush. Although the low velocity removal efficiency is only 25% (suggesting that the high velocity removed four times the amount of material), the extra mass removed during the high velocity flush was still quite small. In study Z, 150 grams are removed during the low velocity flush, and an additional 50 grams are removed during the ensuing high velocity flush. In this case, the low velocity removal efficiency was quite good at 75%, but the high velocity flush still collected an additional 50 grams of material. To some utilities, the benefit of removing this additional 50 grams may offset the costs associated with flushing at the higher velocity.

The end result of these manipulation procedures is to perform the following switch of variables: $(C_i, Q) \rightarrow (M_{i,low}, M_{i,high}) \rightarrow (\hat{M}_i^*, E_i)$, where C_i is the concentration data for parameter i, Q is the flushing rate, $\hat{M}_i^*$ is the normalized theoretical mass of parameter i collected at the high velocity, and E_i is the low velocity mass removal efficiency for parameter i. Viewing the data as normalized theoretical mass and low velocity efficiency allows for comparison to be made within and between different field studies.

Results

Normalized theoretical mass and low velocity efficiency results are provided for total iron, total manganese, and heterotrophic plate count (HPC) in Table 5.7. Only the 14 field trials that met the data validation criteria are included. Total coliform, orthophosphate, total suspended solids, and calcium data were not further analyzed due to limited data availability.

Table 5.7
Normalized data for bulk water trials

Site	Step	$\hat{M}^*_{Fe}$ (mg/ft^2)	E_{Fe} (%)	$\hat{M}^*_{Mn}$ (mg/ft^2)	E_{Mn} (%)	$\hat{M}^*_{HPC}$ (No/ft^2)	E_{HPC} (%)
ANC	5	56.7	69	6.5	88	NA	NA
ANC	13	31.6	66	3.4	42	NA	NA
CLE	5	18.2	53	NA	NA	2.64×10^4	11
DOV	3	13.2	6	0.6	7	2.60×10^6	5
IR	3	12.1	91	NA	NA	NA	NA
IR	8	23.1	42	NA	NA	NA	NA
NNH	3	2,176	11	13.1	8	2.13×10^8	20
NNI	4	23.8	43	2.9	39	1.62×10^7	30
POR	6	798.2	47	35.7	88	8.77×10^6	69
POR	12	346.0	13	14.1	44	9.72×10^6	48
YVR	4	3.8	25	NA	NA	NA	NA
YVR	10	5.5	62	NA	NA	NA	NA
YVW	5	75.3	99	NA	NA	3.51×10^5	96
YVW	13	117.0	99	NA	NA	4.53×10^7	100

No = Number
NA = Not Applicable
$\hat{M}^*_i$ = Normalized Theoretical Mass of Parameter i
E_i = Low Velocity Removal Efficiency for Parameter i

The normalized theoretical mass values for iron and manganese ranged between 3.8-2,176 mg/ft^2 and 0.6-35.7 mg/ft^2, respectively. In each case, iron accumulation was at least an order of magnitude greater than manganese. HPC levels varied from 10^4-10^8 per square foot.

The low velocity removal efficiencies are plotted for each trial in Figure 5.2. These results can be used to help address the question: "Does a higher flushing velocity remove significantly more material?"

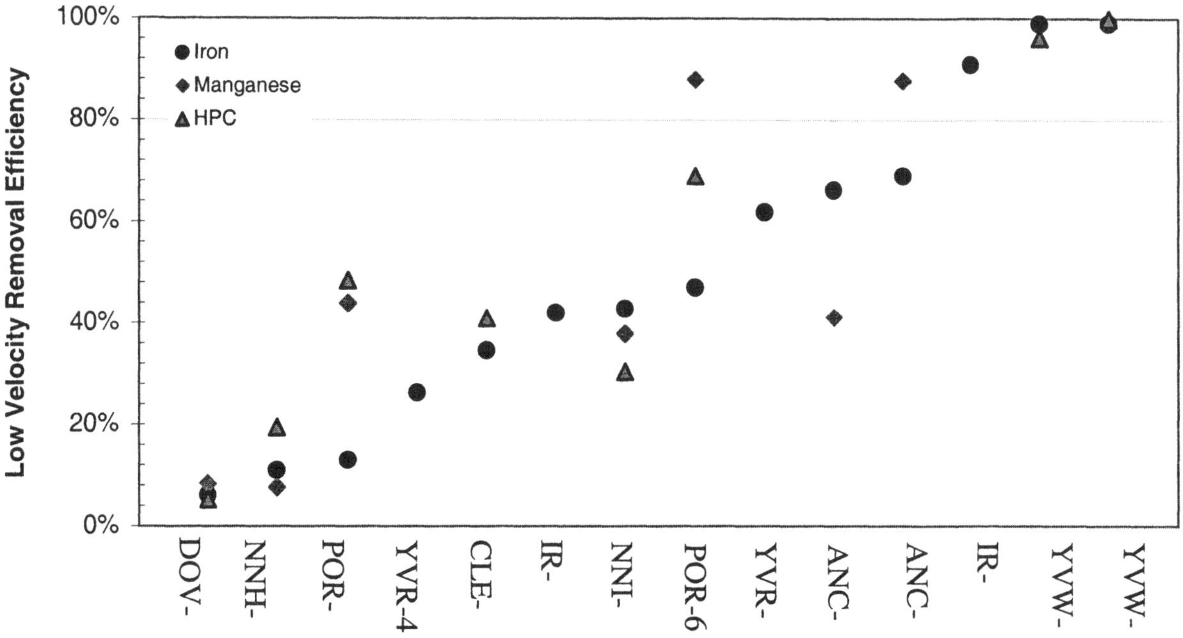

Figure 5.2 Low velocity removal efficiencies for bulk water trials

The low velocity removal efficiencies for iron, manganese, and HPC track each other relatively well. This suggests that none of the contaminants are preferentially removed at the velocities tested. In other words, contaminant-specific flushing velocities for iron, manganese, and HPC appear to be quite close to one another within each system but quite different from each other between systems. Based on these data, discrete contaminant-specific flushing velocities could not be predicted. The low velocity removal data span the range 5-99%, which suggests that universal contaminant-specific flushing velocities may not exist. In other words, the answer to the question: "Does a higher flushing velocity remove significantly more material?" is not as simple as yes or no. Some utilities experienced significant benefit by flushing above 5 fps (1.5 m/s), while others experienced negligible benefit. This phenomenon may be a result of the many variables involved in the field studies, as discussed in the following section.

Variables Affecting Flushing Performance

The normalized mass and removal efficiency results presented in Table 5.7 are a complex function of many physical, chemical, and operational factors. Some of these factors are easily characterized, while others are unknown and/or non-quantifiable. The purpose of this section is to re-analyze the bulk water results from the perspective of some of these variables in order to determine their relative significance.

Primary Variables

The Project Team identified a discrete set of variables suspected of having the *greatest* overall impact on material accumulation, deposit properties, and flushing performance. These variables, referred to as primary variables, are listed below.

- *Time Since Last Flush.* It was originally anticipated that, with all other variables equal, the total amount of accumulated material would be directly related to the time interval between the date of this study and the prior flush. The extent of compaction may also be related to the time since last flushed. If an area had never been flushed, time since installation was used.

- *Previous Flushing Method.* It was originally anticipated that, with all other variables equal, the total amount of accumulated material would be less for those systems that have previously conducted unidirectional flushing compared to dead-end or conventional flushing. Further, those systems that previously conducted conventional flushing may have been unsuccessful in scouring compacted or adherent deposits.

- *Pipe Material.* Pipe material was suspected of having two potential effects on flushing. First, the material may contribute to the mass of specific contaminants, as in the case of corrosion by-products from unlined cast iron mains. Second, the type of pipe may partially dictate the internal surface characteristics (e.g. degree of tuberculation, smoothness coefficient), which could in turn affect flushing removal performance.

- *Pipe Diameter (Size).* Pipe diameter influences flow hydrodynamics as well as boundary shear during flushing. Pipe diameter and flow rate both effect flow velocity. For a given bulk fluid velocity, smaller diameter mains experience higher boundary shear (see Figure 6.2).

- *Source Water and Applied Treatment.* Unfiltered surface waters typically contain more sediment and fines than groundwater. Therefore, it was originally anticipated that unfiltered surface water systems would contain more accumulated and compacted deposits than comparable filtered surface water or groundwater systems. Also, unfiltered groundwater can contain elevated iron and manganese levels compared to surface water.

Secondary Variables

Several other factors could potentially impact flushing effectiveness. These variables, referred to as secondary variables, are much more complicated to evaluate and/or are often not documented. As an example, consider the effect of fire-fighting activities in an area. Fire flows simulate conventional flushing, and may provide scouring and contaminant removal, as well as contaminant dispersal between pipelines. Therefore, if a fire or fire flow testing occurred near any of the test sites since the last documented flush, it would be reasonable to observe lower normalized mass values (ignoring the effect of other variables). However, since the history of hydrant usage is consistently not available at this level of detail, it is very difficult to incorporate this variable into the data analysis. The following is a partial list of secondary variables. For reasons previously discussed, they were not comprehensively included in the data analysis process.

- Hydrant operation (e.g. fire flow, unplanned spot flushing)
- Pipe repair activities
- Routine system hydraulics
- Alternating source waters
- Effects of different corrosion control methods (e.g. phosphate addition, pH adjustment) on corrosion of iron pipe

Iron, Manganese and HPC Results

In order to accurately assess the effect of one variable at a time, it is necessary to hold all other variables constant. However, the large number of variables and unique characteristics of each water system make this task extremely difficult. The number of primary variables is nearly half the number of field studies where bulk water trials were successfully conducted. As a result, there are very few instances where all variables but one are constant between two or more utilities. Three-dimensional bar plots were used to analyze the relative effects of different variables. Normalized theoretical mass results were depicted as a function of primary variables. These graphs were constructed for each parameter (iron, manganese, and HPC) and discussed separately.

Iron Analysis

The normalized theoretical mass results for iron (provided in Table 5.7) are depicted in Figure 5.3. Data bars are numbered for convenience of identification and reference. The geometry of the graph layout suggests that the bars would be expected to grow in size as one moves from the front left corner of the x–y grid to the rear right corner. This trend is somewhat apparent in the graph, although there are some discrepancies. These discrepancies arise because of the influence of the other (labeled) primary and unlabeled secondary variables.

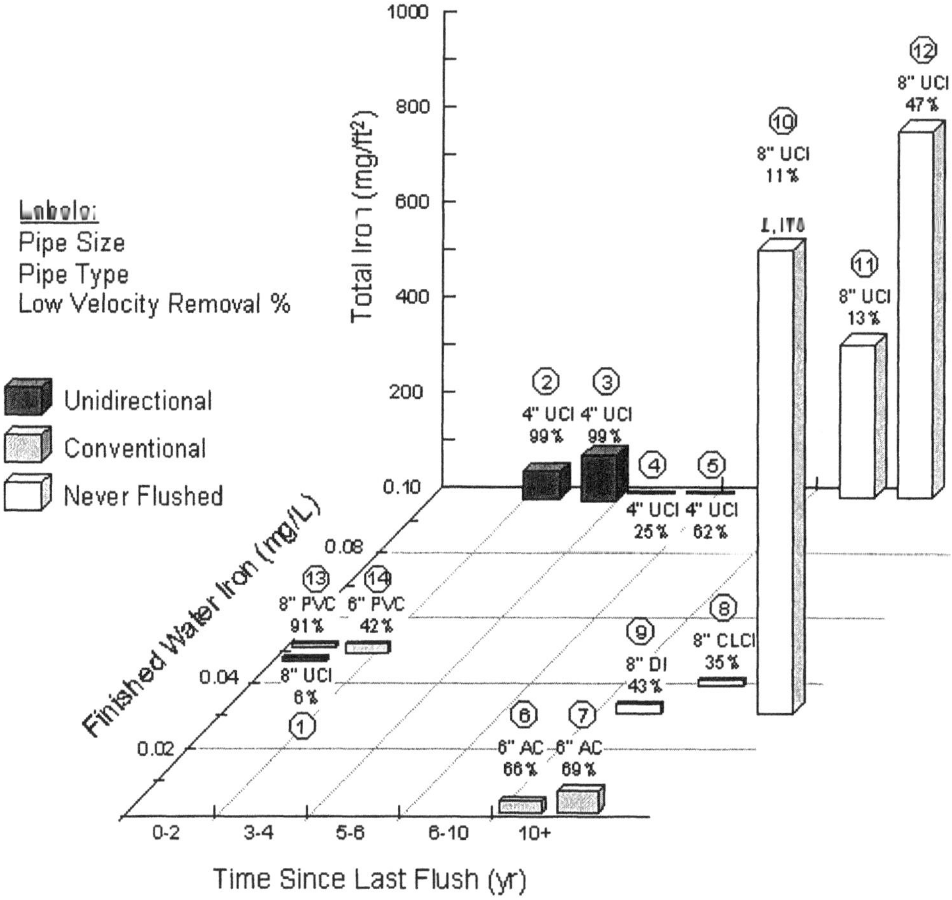

Figure 5.3 Normalized iron results for bulk water trials

Source of Iron. Iron can be introduced into the distribution system either through the finished water leaving the treatment plant (lack of treatment or post-flocculation) or through the corrosion of unlined iron mains. One focus of this study was to determine the relative contribution from these sources, as this has significant implications regarding the ability to remove deposits. Careful comparison of select data bars in Figure 5.3 suggests that pipe material may have a more profound effect on the accumulation of ferric deposits than finished water quality. This conclusion concurs with the findings in the literature review (Chapter 2). Reasonable comparison can be made between points 8, 9, and 10, since all mains are 8-inch diameter and had never been flushed prior to this study. Points 9 and 10 are part of the same distribution system (Newport News) and convey the same finished water. Therefore, the only primary variable different between these two mains is pipe material. The bar for the unlined cast iron pipe (Hilton Village) is more than 91 times taller than that for the ductile iron pipe (Ivy Farms), suggesting that corrosion of unlined cast iron mains provides a significant contribution of iron-rich deposits. This conclusion is reinforced by photographs of pipe sections taken from each part of the distribution system, provided as Figure 5.4.

Hilton Village
Unlined Cast Iron

Ivy Farms
Ductile Iron

Figure 5.4 Pipe sections from the Newport News distribution system

The significance of pipe material on accumulation of iron deposits is further supported by comparison of points 8 and 9. Both points represent 8-inch water mains that were installed roughly 30 years ago. Point 8 consists of cement-lined cast iron pipe while point 9 consists of ductile iron pipe. Based on the previous discussion, neither of these pipe surfaces would be expected to contribute significant corrosion by-products. Therefore, the sole source of accumulated iron would be the finished water. Both pipes convey water with similar iron levels and, as expected, they have accumulated nearly the same amount of iron material since installation. Further, the low velocity removal efficiency was nearly the same for these two mains, and much better than that for point 10. The rough, tuberculated surface of the cast iron pipe used during point 10 may have harbored loose deposits and prevented entrainment at the low velocity. This postulate is consistent with the results of CFD modeling.

Impact of Time on Iron Accumulation. The direct impact of time between flushes on iron accumulation is depicted in Figure 5.5. Variations arise because some systems contained UCI pipes, which resulted in an accelerated accumulation.

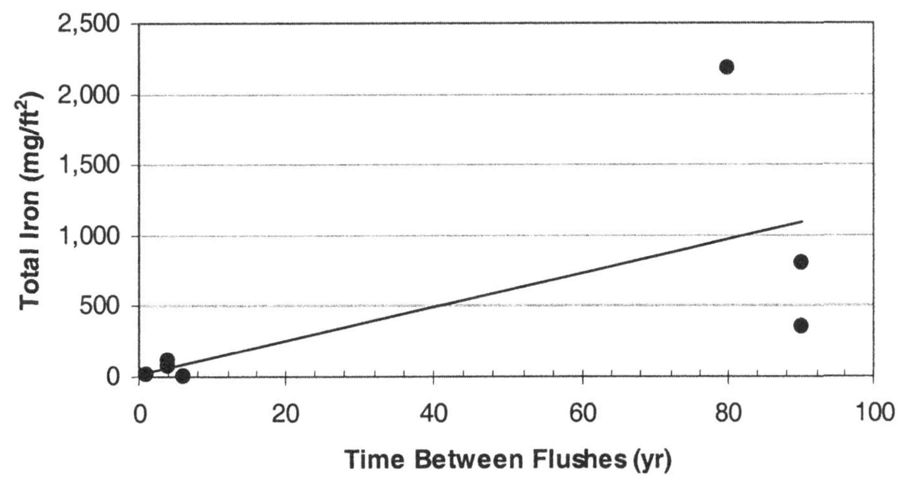

Figure 5.5 Iron accumulation rate for unlined cast iron mains

Manganese Analysis

The normalized theoretical mass results for manganese (provided in Table 5.7) are depicted in Figure 5.6. Data bars are numbered for convenience of identification and reference. The geometry of the graph layout suggests that the bars would be expected to grow in size as one moves from the front left corner of the x–y grid to the rear right corner. This trend generally exists, although it is difficult to observe because of the limited number of data points.

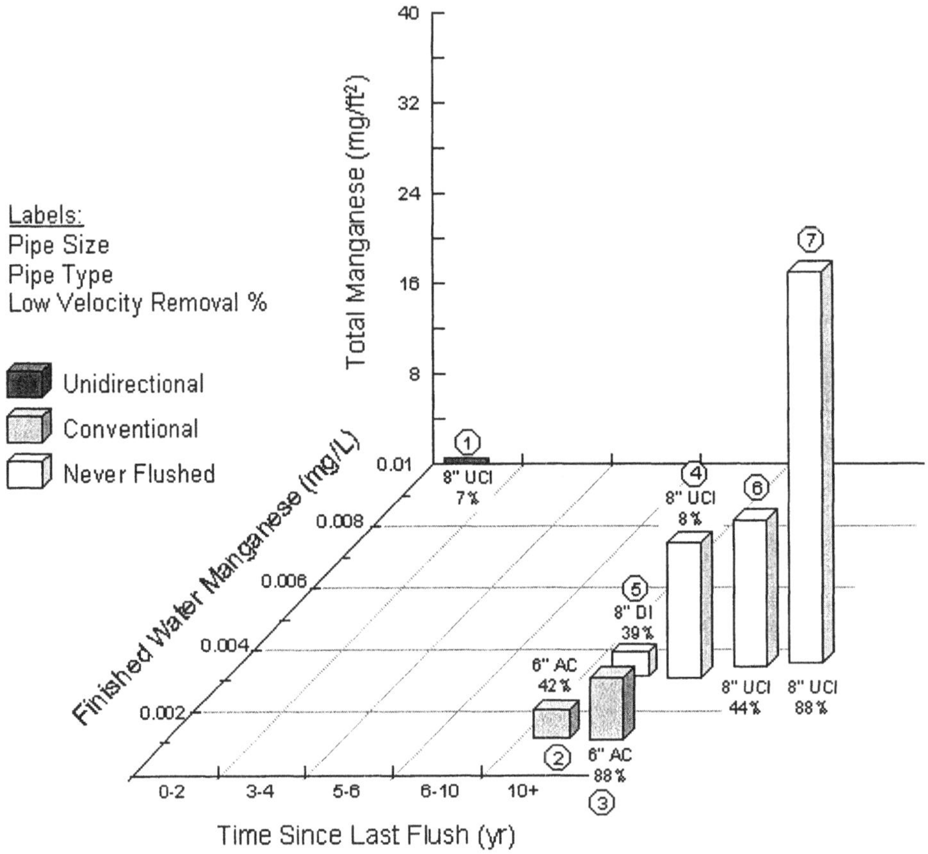

Figure 5.6 Normalized manganese results for bulk water trials

Source of Manganese. Unlike iron, manganese is not a typical constituent of pipe materials. Cast- and ductile-iron pipes contain 0.55% and 0.45% elemental manganese, respectively (AWWA and DVGW-Technologiezentrum Wasser 1996). Asbestos cement and PVC pipes do not contain any manganese. Therefore, pipe material dissolution is not considered to be a significant source of manganese in the distribution system. Rather, it is suspected that most manganese particulates originate from the finished water. However, the data set is too limited to verify this statement. Dover (point 1) contained the highest finished water manganese level (0.01 mg/L) of the valid bulk water trials. However, the accumulated manganese was low at this location because it had been flushed using unidirectional methods within the previous two years. It is unrealistic to compare points 2 and 3 with points 4 – 7 in order to make an assessment about the influence of finished water quality, since the time intervals involved vary significantly.

Impact of Time on Manganese Accumulation. Figure 5.6 provides additional support for the idea that contaminant accumulation is directly related to the time since the previous flush (or the time since installation, if flushing has never been conducted). Points 4 and 5 are part of the same water system, and therefore receive the same finished water. However, the cast iron water mains at Hilton Village (point 4) were installed approximately 50 years earlier than the ductile iron mains installed at Ivy Farms (point 5). As expected, the older mains contain 4.5 times more accumulated manganese than the newer mains.

HPC Analysis

Bacteriological analysis was performed on bulk water samples to better understand the extent of biofilm scouring during unidirectional flushing. Prior studies on biofilm formation have focused on pipe material, internal surface characteristics, and water quality conditions, including disinfectant type and residual. Secondary disinfection with chlorine or chloramines mitigates bacterial regrowth by destroying microorganisms entrained in the bulk water. According to a study by Abernathy and Camper (1998), 2 mg/L of monochloramine was more effective than 1 mg/L of free chlorine at reducing HPC bacteria in both the bulk water and biofilm. Internal corrosion of pipe surfaces creates an environment favorable to bacterial colonization and biofilm formation. As shown through CFD modeling (Chapter 3) pits and cavities provide a surface area largely protected from the shear forces of bulk flow. Also, corrosion by-products can adsorb organic molecules and nutrients from the bulk water that serve as food sources for developing bacterial colonies (Abernathy and Camper 1998). These findings help provide an understanding of how HPC bacteria accumulate and are removed under various hydraulic and water quality conditions.

The normalized theoretical results for HPC bacteria (provided in Table 5.7) are depicted in Figure 5.7. Data bars are numbered for convenience of identification and reference.

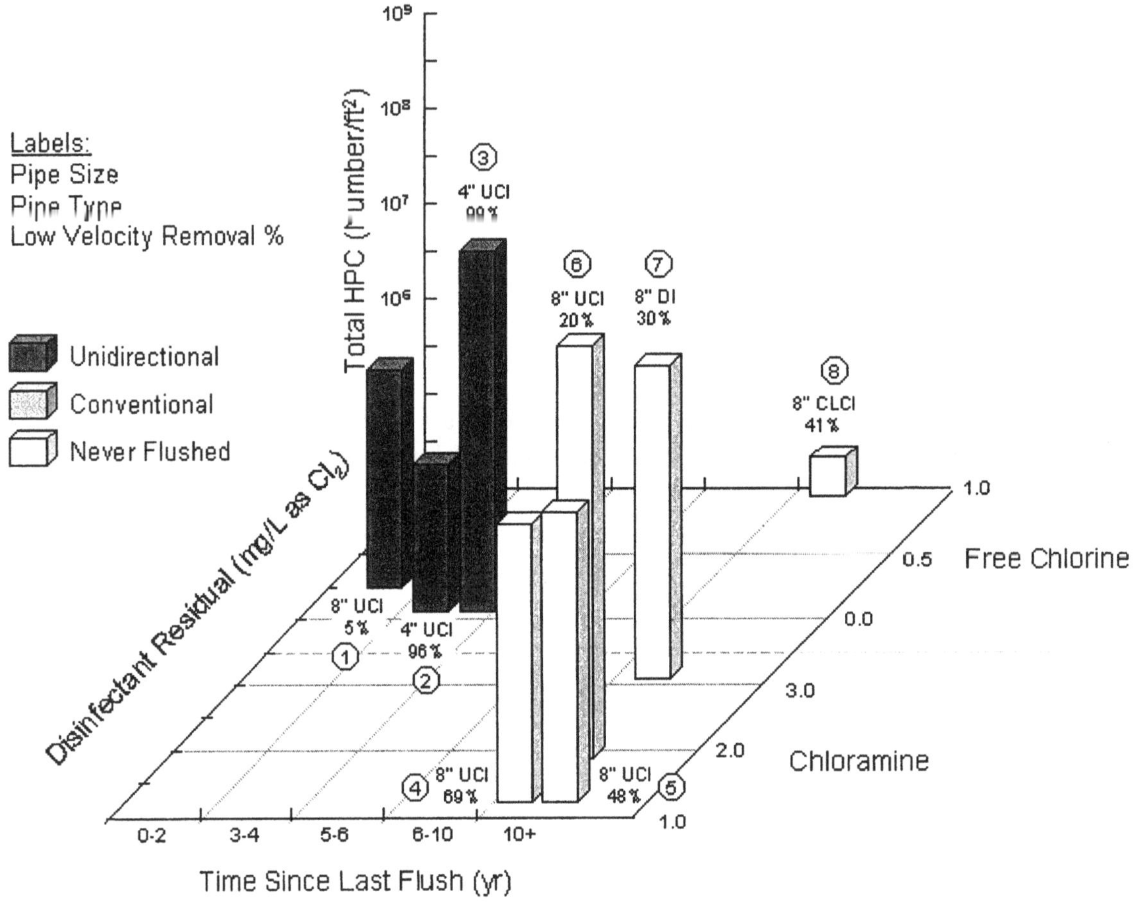

Figure 5.7 Normalized HPC results for bulk water trials

It appears that flushing was effective at scouring some biofilm from distribution mains. However, a determination of the total percentage of biofilm capable of being removed by unidirectional flushing requires a more in-depth analysis of pipe coupons before and after the flush. There is no obvious relationship between disinfectant type and accumulated HPC. Regarding disinfectant concentration, free chlorine levels near 1.0 mg/L produced water with significantly lower bacterial counts than waters with free chlorine levels below 0.5 mg/L.

Low velocity removal efficiencies are also presented in Figure 5.7. The data show no clear relationship between this parameter and either pipe type, time since last flush, or disinfectant residual. However, the low velocity removal efficiency exhibits some relationship to pipe size and the method used for previous flushing (i.e. conventional or unidirectional). In two of the three instances where unidirectional flushing had been conducted within the previous 6 years, the higher velocity provided negligible benefit. In all five instances where flushing had never been previously conducted, the higher velocity proved effective at removing additional HPC (140% more removed, on average).

Velocity Impact

The ability of a specific velocity to shear and entrain deposits is a function of both pipe factors (surface hydraulics) and particle factors (density and adherent properties). However, as discussed in the previous section, the pipe type may largely control the particle properties. Table 5.8 provides a summary of the efficiencies of the low velocity unidirectional flushing events for removing bulk water contaminants.

Table 5.8
Average low velocity removal efficiencies for bulk water trials

Pipe material	E_{Fe} (%)	E_{Mn} (%)
Asbestos cement	67.5%	65.0%
Polyvinyl chloride	66.5%	NT
Ductile iron	43.0%	39.0%
Unlined cast iron[1]	23.7%	46.7%

NT = Not tested
(1) Never been previously flushed

There are two interesting trends apparent in these data. First, low velocity flushing was least effective on unlined cast iron mains which had never been previously flushed. This may be a consequence of the rough nature of corroded surfaces, which contain pockets largely protected from bulk flow shear forces. Second, the iron and manganese removal efficiencies are comparable for both AC and DI mains, but markedly different for unlined cast iron mains. This is likely related to the nature of the deposits in each type of main. For systems composed of AC and DI mains, iron and manganese particulates predominantly originate from treatment breakthrough, post-flocculation or a contribution from cast iron pipe elsewhere in the system. Conversely, for systems composed of unlined cast iron mains, iron appears to be contributed primarily by pipe corrosion while manganese occurrence is related to treatment breakthrough. The nature of the particles formed during the processes of loose deposit compaction and scale formation is considerably different, with the latter being more adherent and difficult to remove at low velocities.

Conclusions from Bulk Water Tests

The results of bulk water sampling and analysis provide some indication of the influence of different variables on deposit accumulation rates and flushing performance. However, since the number of different variables involved is comparable to the size of the data set, only a limited number of comparisons and conclusions can be drawn. These are as follows:

- Pipe material appears to have a greater impact on the accumulation of ferric deposits compared to iron contributed by the finished water.

- The low velocity removal efficiencies for iron, manganese, HPCs are similar to one another within each system, but very different from each other between systems. This suggests that universal contaminant-specific flushing velocities may not exist.
- No obvious relationship was observed between secondary disinfectant type (chlorine vs. chloramine) and accumulated HPCs. However, free chlorine levels near 1.0 mg/L produced water with significantly lower bacterial counts than waters with free chlorine levels below 0.5 mg/L.
- Higher velocities ($\geq$ 5 fps) were effective at removing additional HPC in instances where flushing had not been conducted before. If unidirectional flushing had been conducted within the previous six years, high velocity flushing was of negligible benefit.
- Low velocity flushing was least effective on unlined cast iron mains that had never been flushed.
- Low velocities were more effective at removing accumulated iron deposits in AC and DI mains, compared to unlined cast iron mains.

PARTICLE CAPTURE TRIALS

Particle capture trials were performed as part of the field tests in order to characterize distribution system deposits and evaluate flushing performance. The broader results of particle analyses were used to help answer the following two questions:

(1) “Does a higher flushing velocity remove more material?”
(2) “Does a higher flushing velocity remove different material?”

In total, 16 multi-velocity tests were conducted within 10 different water systems. The process of selecting utilities and specific sections of the distribution system was guided by the intention of capturing a diverse range of water quality conditions, pipe materials, and flushing objectives. Table 5.9 provides a comprehensive summary of particle capture trials and the associated analyses. Each step involved sequential low and high velocity flushing events. Separate analyses were conducted on particles collected from each velocity.

Table 5.9
Testing summary for particle capture trials

Site	Step	Pipe type	Mass Fractionation	Chemical composition	Particle density
ANC	7	AC	×	×	×
BEL	1	AC	×	×	
BEL	2	AC	×	×	
CIN	R1	UCI	×		
CIN	R2	UCI			
CLE	12	CLCI	×	×	×
DOV	11	UCI	×	×	
IR	2	PVC	×		
NNH	3	UCI	×	×	
NNI	8	DI	×	×	
POR	5	UCI	×	×	
SB	8	UCI	×	×	
TH	1	PVC	×	×	×
TH	2	UCI	×	×	×
TH	3	UCI	×	×	×
TH	4	SI	×	×	×

Hydrant nets were used to collect deposits at the point of flow discharge, as illustrated in Figure 5.8. The setup consisted of a 308-micron nylon net enclosed by a 57-micron nylon net. As such, conclusions cannot be drawn for particles smaller than 57 microns in size. The recovered particles were subject to laboratory testing at Montana State University and Thames Water Company. The following analyses were performed to identify key physical and chemical characteristics of distribution system particles:

- Fractionation (Sieving)
- Chemical Composition
- Bacteriological Composition
- Particle Density (Specific Gravity)

Figure 5.8 Hydrant net setup for particle capture trials

Bacteriological composition analysis consisted of measuring the density of HPC bacteria adhered to the recovered particles. However, the results were ultimately discarded. The Project Team concluded that if the bacteria were capable of being removed from the pipe surface, they were equally capable of being scoured from the particles as the water violently diffused through the nets. Further, the pore size of the fine net (57-micron) would permit the passage of several types of bacteria and silt particles capable of harboring bacteria.

Data Validation

The conditions of each study were carefully evaluated to determine the appropriateness of using the results to address the issues of particle characterization and velocity effectiveness. Several data validation criteria were established in order to eliminate compromised data from further analysis. The particle capture tests were required to satisfy each of the criteria listed in Table 5.10 in order to be included in subsequent data analysis related to velocity effectiveness.

This validation process was performed to ensure data quality and pending conclusions. Despite utility efforts to conform to these criteria, only 5 trials were completely successful. The process of conducting a particle capture test was substantially more intensive than routine hydrant flushing. Net maintenance and pitot pressure measurements were given foremost attention, but dechlorination and diffusion were also required. The demand of these requirements contributed to the pool of difficulties experienced during many of the testing events.

Table 5.10
Data validation criteria for particle capture trials

1. Flushing must be unidirectional and in accordance with the protocol.
2. A minimum of one pipe volume must be flushed at each velocity.
3. The main must be flushed to clear (turbidity less than 5 NTU) at each velocity.
4. The hydrant nets must remain in-place during the entire duration of the flush.
5. The nets must retain structural integrity throughout the entire flush.
6. The low velocity flush must reside in the range 2.0 – 4.5 fps (0.6 – 1.4 m/s).
7. The high velocity flush must reside in the range 5.0 – 9.0 fps (1.5 – 2.7 m/s).
8. The low and high velocity tests must differ by at least 2.0 fps (0.6 m/s).
9. The entire length of main must be of the same nominal diameter, excluding the effects of diameter reduction due to accumulated materials.

In some cases, the data from invalidated trials are still meaningful and are included in the analyses. This is generally limited to particle characterization data, including chemical composition and density.

Data Manipulation

Particle mass data were manipulated to account for the effects of different main sizes and lengths. Equations 5-3 through 5-6 (previously defined) were applied. The end result of these manipulation procedures is to perform the following switch of variables: $(M_{Total,low}, M_{Total,high}) \rightarrow (\hat{M}^*_{Total}, E_{Total})$, where $M_{Total,\,j}$ is the total particle mass at velocity j, $\hat{M}^*_{Total}$ is the normalized theoretical total particle mass collected at the high velocity, and E_{Total} is the low velocity particle mass removal efficiency.

Results

Results are presented and discussed for mass collected, particle size classification, chemical composition, and particle density.

Mass Collected

Table 5.11 provides a summary of the mass results for the 5 utilities where data met the validation criteria. Converted data are provided in Table 5.12. The theoretical total mass values vary significantly, largely because each test involved mains of different diameters and lengths. In order to make comparisons between different utilities, mass data were normalized to the total internal surface area of the main being flushed. The results are provided in Table 5.13. Low velocity removal efficiencies varied between 0.0% and 44.5%.

Table 5.11
Mass and velocity data for particle capture trials

Site	Step	Low velocity fps (m/s)	$M_{Total,low}$ (grams)	High velocity fps (m/s)	$M_{Total,high}$ (grams)
ANC	7	3.8 (1.2)	51.0	6.0 (1.8)	143.5
BEL	2	4.0 (1.2)	79.9	7.6 (2.3)	99.8
CLE	12	2.9 (0.9)	11.0	5.9 (1.8)	20.5
NNI	8	2.8 (0.9)	147.7	5.6 (1.7)	457.8
POR	5	3.4 (1.0)	T	6.1 (1.9)	75.8

T = Trace amount of mass collected

Table 5.12
Converted mass data for particle capture trials

Site	Step	M^*_{Total} (grams)	E_{Total} (%)
ANC	7	194.5	26
BEL	2	179.7	44
CLE	12	41.2	29
NNI	8	605.5	24
POR	5	75.8	0

Table 5.13
Normalized data for particle capture trials

Site	Step	$\hat{M}^*_{Total}$ (mg/ft^2)	E_{Total} (%)
ANC	7	66.3	26
BEL	2	190.6	44
CLE	12	8.4	29
NNI	8	199.4	24
POR	5	20.9	0

The normalized mass results still exhibit a substantial degree of variability. This is a direct consequence of the fact that normalized mass is a complex function of many other factors. The same primary and secondary variables previously identified are applicable.

In order to accurately assess the effect of one variable at a time, it is necessary to hold all other variables constant. However, the large number of variables and unique characteristics of each water system complicate this procedure. In this portion of the study, the number of primary variables (6) exceeded the number of valid particle data sets (5), thus making it very difficult to compare the role of different variables.

Particle Size Classification

Particles were washed from the hydrant nets into 1-liter beakers using sterile water. Each net pair was washed into a separate beaker. The beakers were then incubated at 65 °C for 2-10 days to remove excess water. Finally, the dried particles were fractionated via sieving. The following bin sizes were employed: No. 4 (≥ 4.75 mm), No. 8 (2.36 mm), No. 16 (1.18 mm), No. 30 (600 μm), No. 50 (300 μm), No. 100 (150 μm), and No. 200 (≤ 75 μm). To differentiate the myriad of particle sizes collected in hydrant nets, the sieve ranges were converted to phi values (ϕ_k) according to equation 5-7.

$$\phi_k = -\log_2(d) \qquad (5\text{-}7)$$

Where:
ϕ_k = Phi value for sieve bin k
d = Minimum size opening of sieve bin k (mm)

Table 5.14 provides a portion of the phi scale, while Figure 5.9 provides phi-specific particle characterization (Wentworth 1922).

Table 5.14
Grain size classification

Size (mm)	4	2	1	0.5	0.25	0.125	0.063
Phi value	-2	-1	0	1	2	3	4

Source: Wentworth 1922

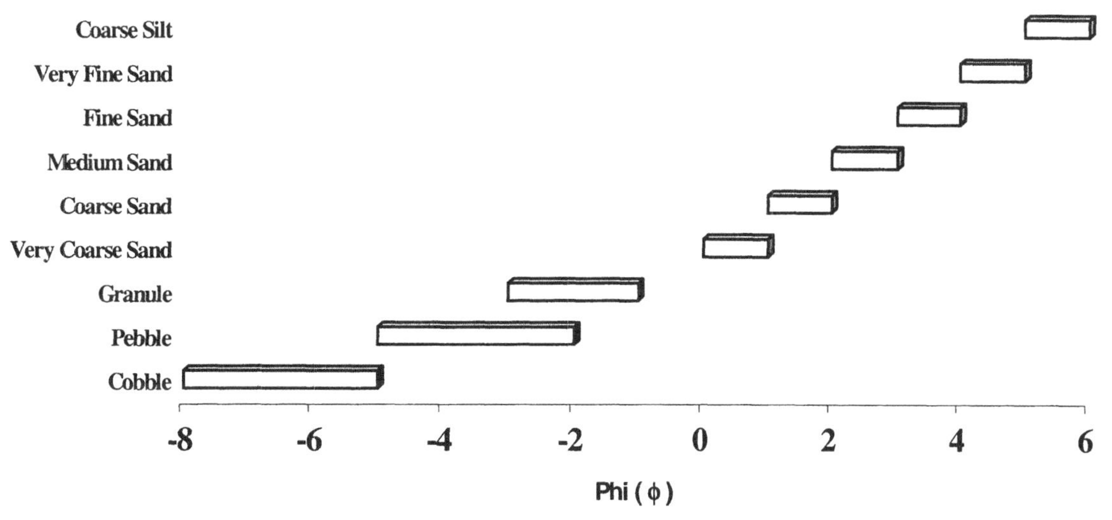

Figure 5.9 Wentworth particle size classification

Particle size analysis can be used to help answer the question: "Does a higher flushing velocity remove material with different physical properties?" Particle size histograms were generated for the five particle trials that met data validation criteria. Three histograms are provided for each trial: one for the low velocity, one for the ensuing high velocity, and one for the theoretical high velocity (low plus high, indicated with an asterisk). The histograms depict the mass percentage of deposits in each sieve bin (converted to ϕ) for a specific velocity. As such, the area under each curve is 100%.

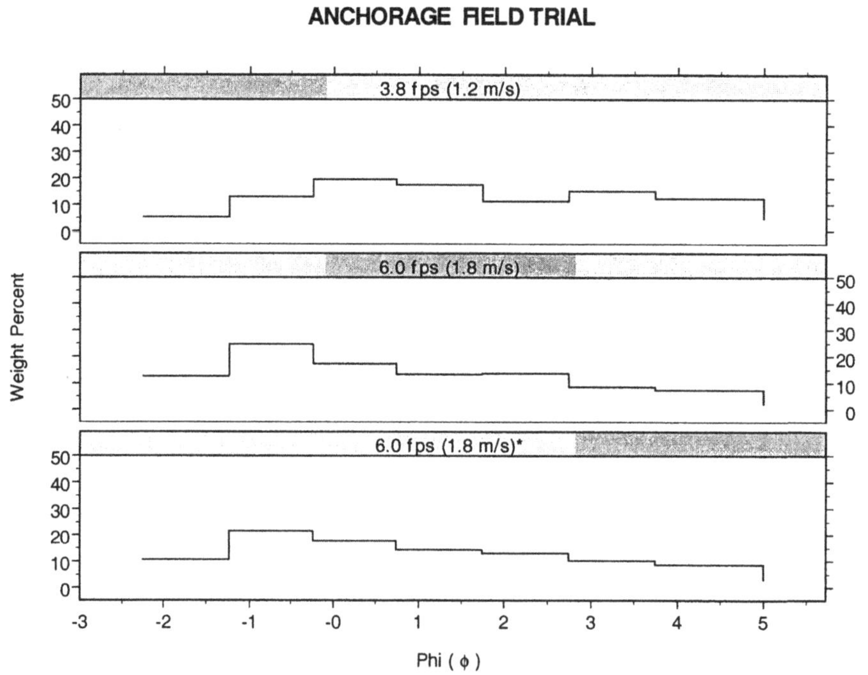

Figure 5.10 Particle size histogram for Anchorage

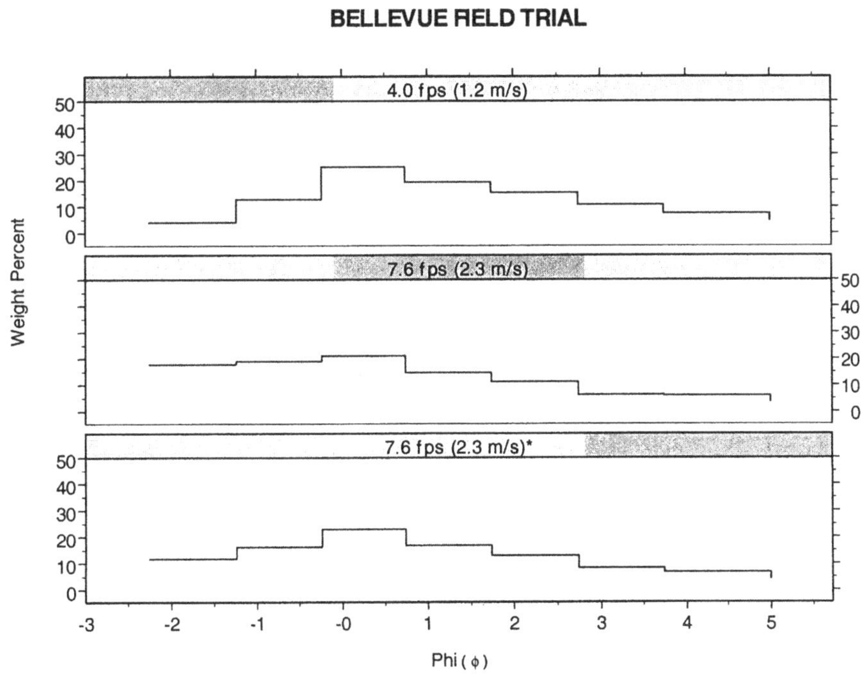

Figure 5.11 Particle size histogram for Bellevue

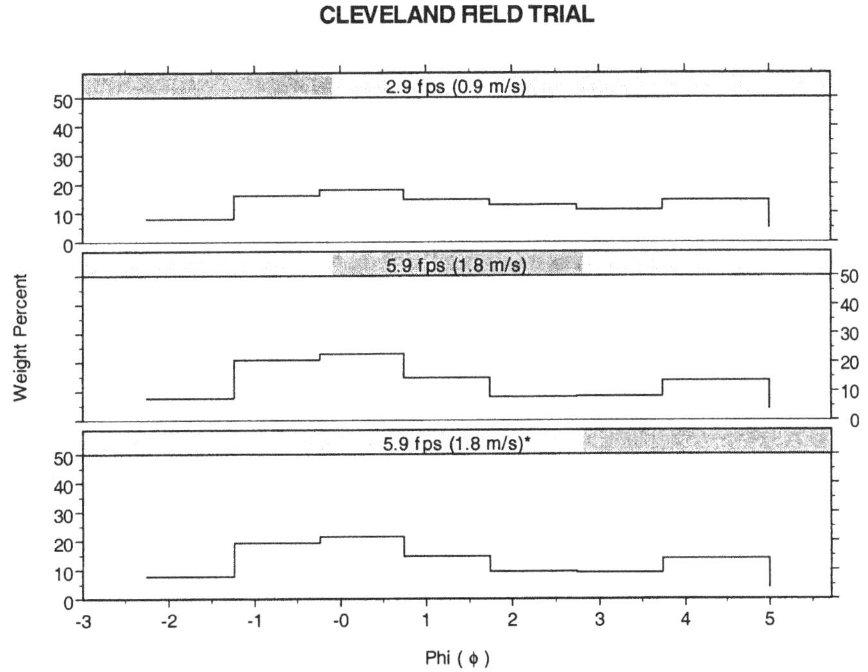

Figure 5.12 Particle size histogram for Cleveland

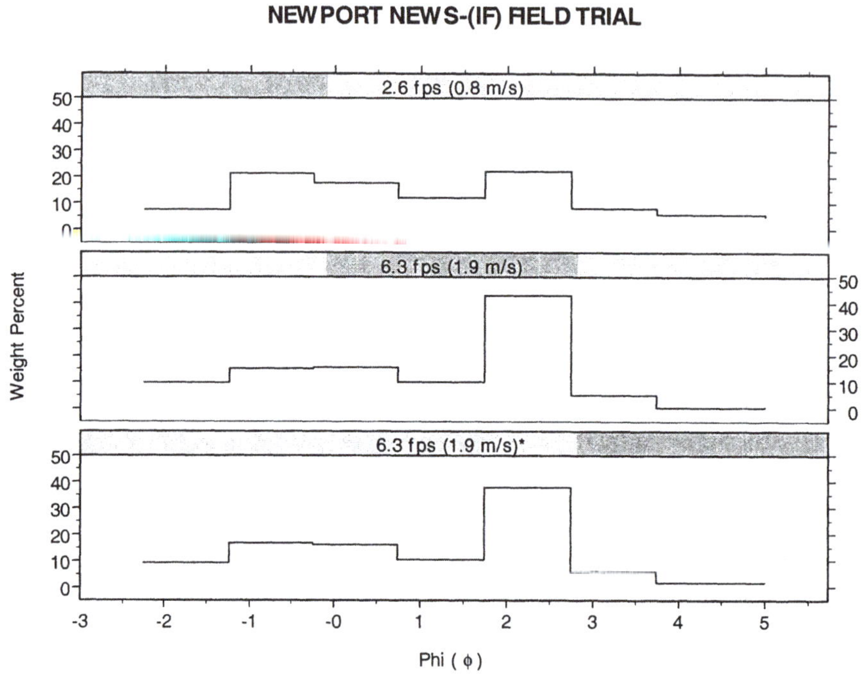

Figure 5.13 Particle size histogram for Newport News

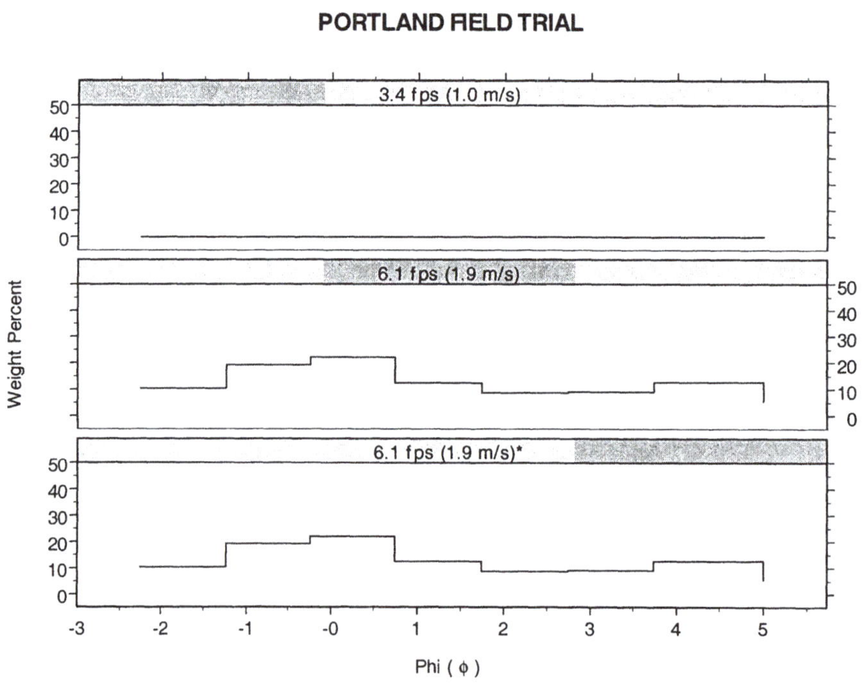

Figure 5.14 Particle size histogram for Portland

The low and high velocity histograms were compared to evaluate particle size differences within and between each water system. Based on visual inspection of the particle size histograms, the low and high velocity results are similar for the Cleveland and Newport News (Ivy Farm) field studies. This suggests that, *within* these systems, similar size particles are removed regardless of the flushing velocity. However, the histograms differ significantly *between* these two systems. With respect to the Bellevue and Anchorage field studies, the particle histograms for the low and high velocity flushes are distinctly different. The high velocity results are skewed to negative phi values, corresponding to larger deposits. In each case, high velocity flushing removed additional and different sized material relative to the low velocity flush.

Sieve data were subject to the following statistical analyses in order to quantify the differences in particle sizes removed during low and high velocity flushes. Both variables are defined on a phi basis.

- Mean (μ_j): Average particle size at velocity *j*.
- Standard Deviation (σ_j): Standard deviation of the particle sizes collected at velocity *j*.

$$\mu_j = \frac{\phi_{16} + \phi_{50} + \phi_{84}}{3} \tag{5-8}$$

$$\sigma_j = \frac{\phi_{84} - \phi_{16}}{4} + \frac{\phi_{95} - \phi_5}{6.6} \tag{5-9}$$

The notation ϕ_x indicates the phi value corresponding to the xth percentile when cumulative weight percent is plotted on a probability ordinate scale, as depicted in Figure 5.15. Linear interpolation was used to estimate phi values.

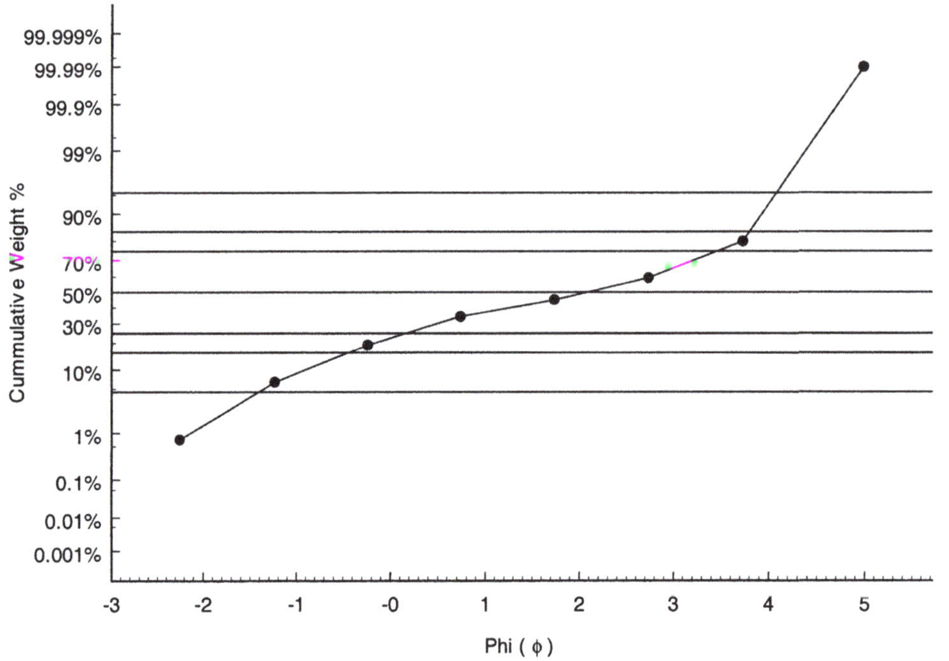

Figure 5.15 Sample probability-normal plot of cumulative weight percent

Standard descriptions of the particles associated with various μ_j and σ_j values are provided in Table 5.15 and Table 5.16, respectively. Statistical results from the five valid particle capture trials are summarized in Table 5.17.

Table 5.15
Description of mean values

Mean (μ_j)		Result
Phi	Normal (mm)	
-∞ – -1	2 – ∞	Gravel
-1 – 0	1 – 2	Very coarse sand
0 – 1	0.5 – 1	Coarse sand
1 – 2	0.25 – 0.5	Medium sand
2 – 3	0.125 – 0.25	Fine sand
3 – 4	0.0625 – 0.125	Very fine sand
4 – 8	0.0039 – 0.0625	Silt
8 – ∞	-∞ – 0.0039	Clay

Source: Adapted from Wentworth (1922)

Table 5.16
Description of standard deviation values

Standard deviation (σ_j)	Result
0 – 0.35	Very well sorted
0.35 – 0.5	Well sorted
0.5 – 0.71	Moderately well sorted
0.71 – 1	Moderately sorted
1 – 2	Poorly sorted
2 – 4	Very poorly sorted
4 – ∞	Extremely poorly sorted

Source: Adapted from Folk and Ward (1957)

Table 5.17
Statistical description of flushed pipe deposits

Anchorage						
Parameter	3.8 fps (1.2 m/s)		6.0 fps (1.8 m/s)		6.0 fps* (1.8 m/s*)	
	Value	Description	Value	Description	Value	Description
Mean	0.63	Coarse sand	-0.18	Very coarse sand	0.02	Coarse sand
Std. dev.	1.98	Poorly sorted	1.93	Poorly sorted	1.96	Poorly sorted
Bellevue						
Parameter	4.0 fps (1.2 m/s)		4.7+ 7.6 fps (1.4 + 2.3 m/s)		7.6 fps* (2.3 m/s*)	
	Value	Description	Value	Description	Value	Description
Mean	0.37	Coarse sand	-0.40	Very coarse sand	-0.05	Very coarse sand
Std. dev.	1.79	Poorly sorted	2.10	Very poorly sorted	1.99	Poorly sorted
Cleveland						
Parameter	3.0 fps (0.9 m/s)		5.9 fps (1.8 m/s)		5.9 fps* (1.8 m/s)	
	Value	Description	Value	Description	Value	Description
Mean	0.48	Coarse sand	0.27	Coarse sand	0.32	Coarse sand
Std. dev.	2.08	Very poorly sorted	2.05	Very poorly sorted	2.08	Very poorly sorted
Newport News (Ivy Farm)						
Parameter	2.6 fps (0.8 m/s)		6.6 fps (2.0 m/s)		6.6 fps* (2.0 m/s)	
	Value	Description	Value	Description	Value	Description
Mean	0.10	Coarse sand	0.13	Coarse Sand	0.08	Coarse Sand
Std. dev.	1.45	Poorly sorted	1.08	Poorly Sorted	1.6	Poorly Sorted
Portland						
Parameter	3.4 fps (1.0 m/s)		6.1 fps (1.9 m/s)		6.1 fps* (1.9 m/s*)	
	Value	Description	Value	Description	Value	Description
Mean	T	T	0.20	Coarse sand	0.20	Coarse sand
Std. dev.	T	T	2.2	Very poorly sorted	2.2	Very poorly sorted

T = Trace amount of mass collected
* = Theoretical results

Based on the values of mean particle sizes, most distribution systems deposits are classified as coarse or very coarse sand. On average, the particles removed during the high velocity flushes were larger in size. As discussed previously, the largest discrepancies between the low and high velocity results occurred for the Anchorage and Bellevue field studies. In these studies, the average particle removed at the high velocity was 1.7-1.8 times larger than that removed at the low velocity. Although both of these trials involved AC pipe, there are insufficient data to infer a relationship between particle size distribution and pipe material. Further, it is important to note that this physical particle difference does not necessarily represent a chemical difference.

Chemical Composition

Chemical composition analyses were conducted on dried particles. This study did not focus on identifying the exact chemical formulae (e.g., Fe_xO_y) of deposits collected in the hydrant nets. Rather, percent mass composition was determined for the following elements: iron, calcium, aluminum, and/or manganese. The results were used to address the following three questions:

(1) What is the composition of distribution system particles?
(2) Is there a relationship between particle composition and pipe material or source water quality?
(3) Do the particles removed at low and high velocities differ in their chemical make-up?

Since the first two questions are related to material characterization and are not velocity-specific, previously invalidated data could be considered. Figure 5.16 portrays the composition results for the 13 field studies where elemental analyses were conducted. In some cases, certain parameters were not analyzed and there is no corresponding data bar. In all cases, at least 40% of the particle composition was composed of one or more unidentified element(s), most likely including carbon, oxygen, and silica.

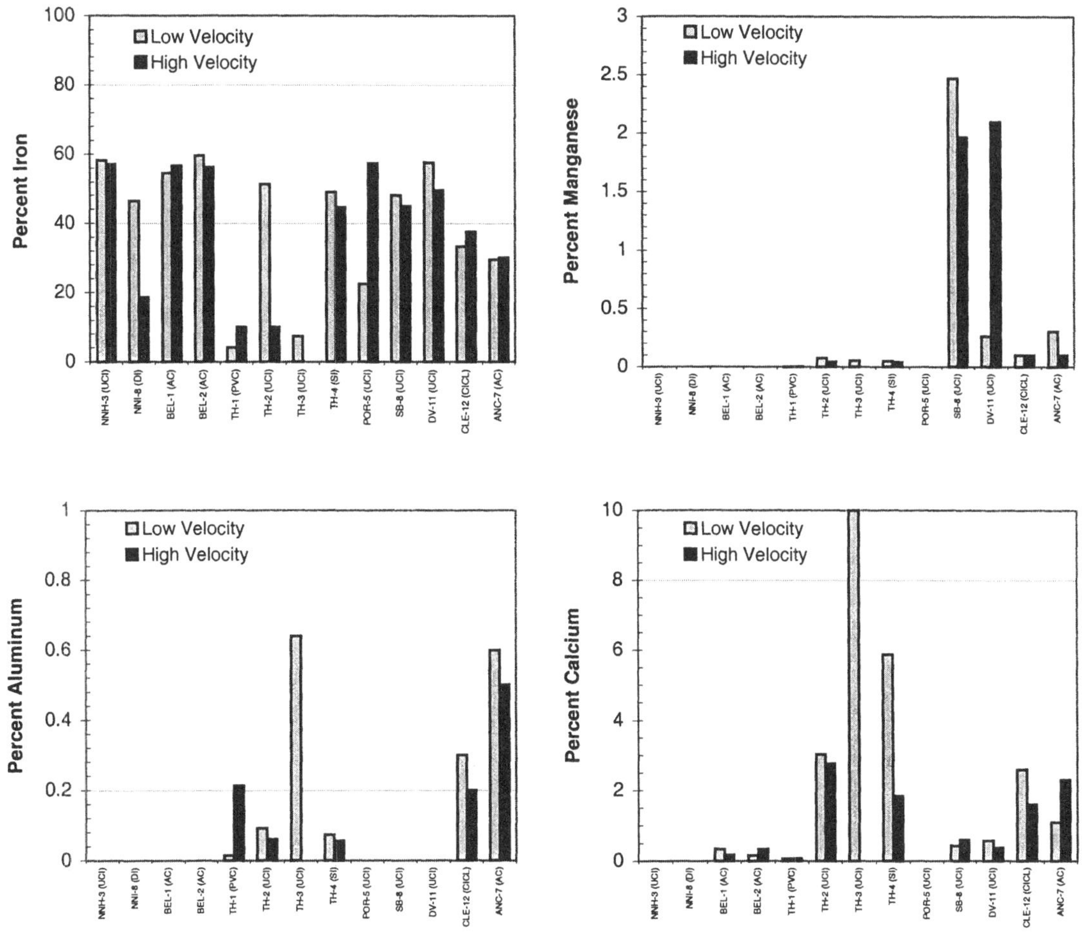

Figure 5.16 Elemental composition of distribution system particles

Iron compositions varied widely (4-60%), but in all cases the iron fraction was one or more orders-of-magnitude greater than the manganese, aluminum, or calcium fractions. In general, field tests involving UCI main produced particles with the highest iron composition. This observation is consistent with the proposal that iron deposits are influenced by pipe material significantly more than finished water, as discussed in the bulk water results. The particle compositions observed in samples from NNI-8, CLE-12, and ANC-7, all of which employed non-UCI pipe, may have been influenced by particles from nearby UCI mains.

Manganese fractions were significantly higher in DOV-11 and SB-8 relative to the other utilities, although they still comprised less than 3% of the total particle mass. Since manganese is not a significant element of any pipe materials, this observation is related to the fact that these two systems purvey groundwater with moderate levels of manganese.

Calcium fractions varied between 0 and 10%. Table 5.18 compares the finished water calcium levels with the calcium fraction in particles. These data suggest that source water calcium contribution dominates AC pipe contribution.

Table 5.18
Select calcium data

Site	Step	Pipe type	Finished water calcium[1]	Particle calcium fraction
ANC	7	AC	28 mg/L	1.1 – 2.3 %
BEL	1	AC	14 mg/L	0.2 – 0.4 %
BEL	2	AC	14 mg/L	0.2 – 0.3 %
CLE	12	CLCI	33 mg/L	1.6 – 2.6 %

(1) Average value

In order to address whether the particles removed at low and high velocities differ, iron results from the five valid data sets were compared (Table 5.19). Since chemical formulae were not determined, similarity in particle iron compositions is a necessary but not fully sufficient condition for chemical resemblance.

Table 5.19
Iron fraction of distribution system deposits

Site	Step	Velocity fps (m/s)	Particle iron fraction
ANC	7	3.8 (1.2)	29.6 %
		6.0 (1.8)	30.2 %
BEL	2	4.0 (1.2)	59.7 %
		7.6 (2.3)	56.3 %
CLE	12	3.0 (0.9)	33.3 %
		5.9 (1.8)	37.6 %
NNI	8	2.6 (0.8)	46.4 %
		6.6 (2.0)	18.6 %
POR	5	3.4 (1.0)	22.6%
		6.1 (1.9)	57.4 %

The particles removed during the high velocity flushes for NNI-8 and POR-5 were considerably different from those removed at the low velocity. For NNI-8, the iron fraction dropped 60% between the low and high velocity flushes. Conversely, for POR-5 the iron fraction increased by 154% between the low and high velocity flushes. While the rationale for these changes is not fully understood, they do suggest that high velocity flushing removed particles of a different chemical nature. The particles removed during the low and high velocity flushes for ANC-7, BEL-2, and CLE-12 appear to be similar in chemical nature.

Particle Density

The CFD modeling results indicate that particle density is a critical variable for determining whether a loose deposit can be removed under various hydraulic conditions. Particle density is a function not only of elemental composition, but also the structural arrangement of the constituent molecules. It is possible for two particles of dissimilar composition to have very similar specific gravities. Also, it is possible for two particles of similar composition to have different specific gravities. Because particle composition analysis focused on elemental mass fraction, and in most cases nearly 50% of the mass was unidentified, it is difficult to identify relationships between particle composition and density. However, the role of water quality and pipe conditions in generating particles of a specific nature and density can be speculated.

Specific gravity measurements were performed with a pyknometer at Thames Water Company according to BS1377 (Part 2: 1990. Determination of Particle Density- small pyknometer method). Analyses were conducted on particles from six different multi-velocity field tests, four of which were invalidated. The results were used to address the following three questions:

(1) What is the range of specific gravities for common distribution system deposits?
(2) Is there a relationship between specific gravity and pipe type or source water quality?
(3) Is the specific gravity of particles removed at low and high velocities different?

Since the first two questions are related to material characterization and are not velocity-specific, previously invalidated data could be considered. The complete set of density results are provided in Table 5.20 and portrayed in Figure 5.17. Mass-weighted average specific gravity was calculated according to equation 5-10.

$$\bar{S}_j = \frac{\sum S_{k,j} \times M_{k,j}}{\sum M_{k,j}} \quad (5\text{-}10)$$

Where:

$\bar{S}_j$ = Mass-weighted average specific gravity of particles collected at velocity j

$S_{k,j}$ = Specific gravity of particles collected in sieve bin k at velocity j

$M_{k,j}$ = Mass of particles collected in sieve bin k at velocity j

Table 5.20
Specific gravity results

Site	Step	Pipe type	Specific gravity[1]	
			Low velocity	High velocity
ANC	7	AC	2.39	2.94
CLE	12	CICL	3.10	2.81
TH[2]	1	PVC	3.12	3.01
TH[2]	2	UCI	2.94	2.87
TH[2]	3	UCI	2.72	NA
TH[2]	4	SI	3.15	2.96

NA = Not Applicable
(1) Mass-weighted average
(2) Not unidirectional or velocity-specific

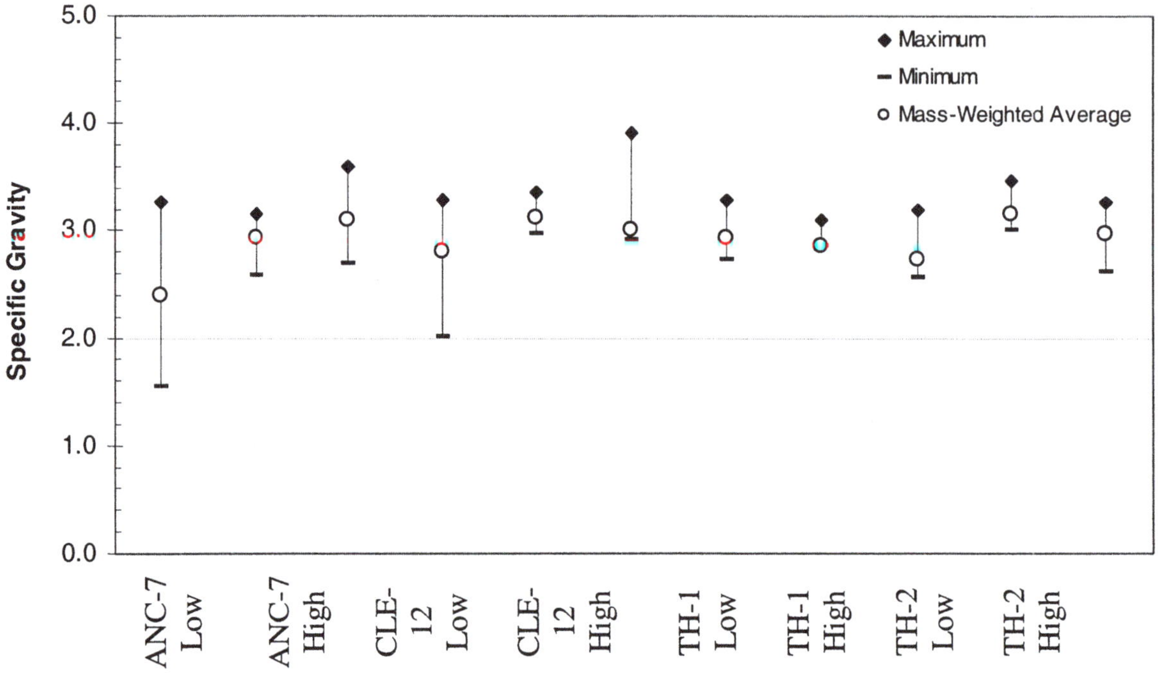

Figure 5.17 Specific gravities of distribution system particles

Similar particle densities were observed for all 6 trials and 5 pipe types. Mass-weighted average specific gravities fell within the range 2.39 – 3.15. Overall, the minimum and maximum specific gravities observed were 1.56 and 3.92, respectively. Based strictly on CFD modeling of loose deposits, the particles removed during the high velocity flush should have a higher average specific gravity than those removed during the low velocity flush. However, this relationship does not exist in most cases, because the assumption of primarily loose deposits within distribution systems is invalid. In real distribution mains, deposits can be loose, compacted, cohesive, or adherent, and this differentiation cannot be made solely based on the material removed during flushing. The significance of particle density is greatly reduced if the deposits exhibit some adherence to the underlying surface, or are compacted under particles of greater density.

The specific gravities of several common distribution system precipitates are provided in Table 5.21. The density ranges for these particles overlap well with those observed during the field tests, suggesting that these and similar precipitates are present in the distribution systems for each of the field trials.

Table 5.21
Specific gravities of select precipitates

Name	Formula	Specific gravity
Gypsum	$CaSO_4 \cdot 2H_2O$	2.30 – 2.37
Aluminosilicates	Various	2.50 – 2.70
Calcite	$CaCO_3$	2.71
Dolomite	$Ca \cdot Mg(CO_3)_2$	2.86
Goethite	α-FeOOH	3.30 – 4.30
Siderite	$FeCO_3$	3.96
Manganese dioxide	MnO_2	5.04 – 5.08
Magnetite	Fe_3O_4	5.20 – 5.30

Source: Data from AWWA and DVGW-Technologiezentrum Wasser 1996

Figure 5.18 illustrates the relationship between specific gravity and particle composition, based on data from the CLE-12 and ANC-7 tests. The data do not point to any obvious relationships between these variables. Specific gravity may be further influenced by unidentified element(s).

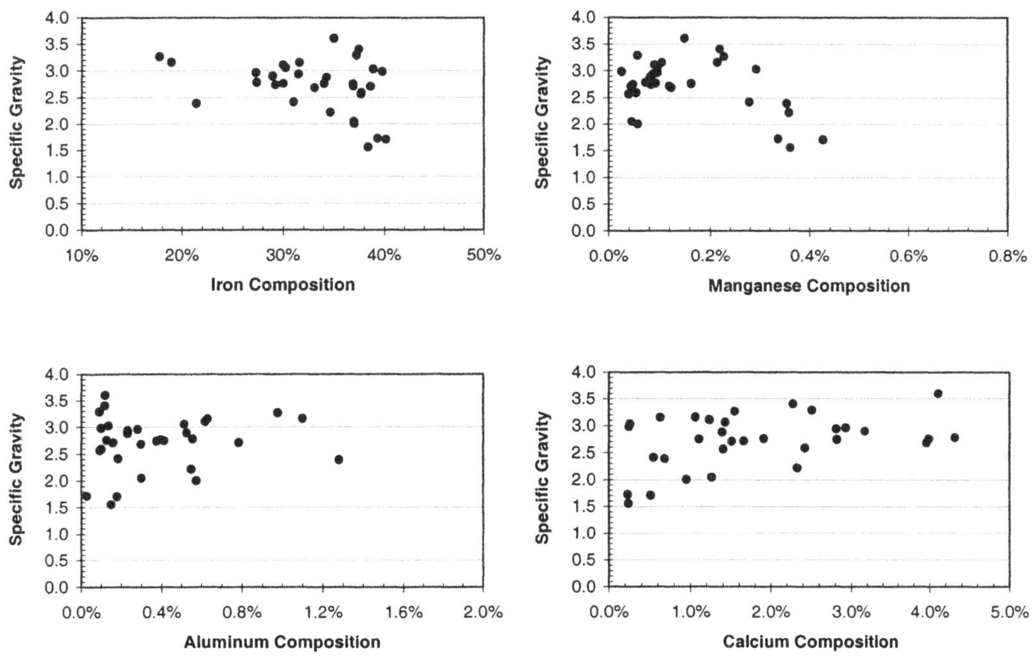

Figure 5.18 Relationship between particle specific gravity and elemental composition

Conclusions from Particle Capture Tests

Particle analyses were performed to address issues related to materials characterization and flushing performance. It should be noted that conclusions could not be drawn for particles smaller than 57 microns due to the hydrant net sizes used in this study. The broader results of particle analyses were used to help answer the following two questions:

(1) "Does a higher flushing velocity remove more material?"
(2) "Does a higher flushing velocity remove different material?"

Since the chemical formulae of distribution system deposits were not fully determined, particle size, elemental composition, and specific gravity were used to assess particle differences. Equations 5-11 through 5-14 were defined to establish quantitative answers to the preceding questions. The results are provided in Table 5.22. These data were then used to generate Table 5.23, which provides qualitative responses to the above two questions. These results can be used by utilities to determine whether the benefits of high velocity flushing outweigh the costs.

$$\text{Difference in mean particle size} = 2^{-\mu_{High}} - 2^{-\mu_{Low}} \tag{5-11}$$

$$\text{Ratio of mean particle size} = \frac{2^{-\mu_{High}}}{2^{-\mu_{Low}}} \tag{5-12}$$

$$\text{Ratio of mean iron fraction} = \frac{Fe_{High}}{Fe_{Low}} \tag{5-13}$$

$$\text{Difference in mean specific gravity} = \bar{S}_{high} - \bar{S}_{low} \tag{5-14}$$

Where:
μ_j = Phi-based average size of particle collected at velocity j
Fe_j = Average iron fraction of particles collected at velocity j (%)
$\bar{S}_j$ = Mass-weighted average specific gravity of particles collected at velocity j

Table 5.22
Quantitative evaluation of particle results

Site	Step	Pipe size	Low velocity removal efficiency	Difference in mean particle size (mm)	Ratio of mean particle size	Ratio of mean iron fraction	Difference in mean specific gravity
ANC	7	6-in (150-mm)	26%	0.49	1.75	1.02	0.55
BEL	2	6-in (150-mm)	44%	0.55	1.71	0.94	ND
CLE	12	8-in (200-mm)	29%	0.11	1.16	1.13	-0.29
NNI	8	8-in (200-mm)	24%	-0.02	0.98	0.40	ND
POR	5	8-in (200-mm)	0%	NA	NA	2.54	ND

NA = Test could not be applied
ND = No Data

Table 5.23
Qualitative evaluation of particle results

Site	Step	Total mass	Average particle size	Elemental composition	Specific gravity
ANC	7	SI	SI	NC	MI
BEL	2	MI	SI	NC	NA
CLE	12	SI	MI	NC	MD
NNI	8	SI	NC	SD	NA
POR	5	SI	NA	SI	NA

SI = Substantial increase from low to high velocity
MI = Moderate increase from low to high velocity
NC = Negligible change from low to high velocity
MD = Moderate decrease from low to high velocity
SD = Substantial decrease from low to high velocity
NA = Not Applicable

High velocity flushing removed substantially more material in 4 of the 5 field trials. However, based on size, composition, and density data, it appears as though the contaminants removed at low and high velocities were not always different. Perhaps the most important finding from the field trials is that "real-world" deposits can be loose, compacted, cohesive, or adherent, and these properties dictate the ability to scour and entrain specific particles. Thus, the development of universal contaminant-specific velocities may not be a worthwhile pursuit. Rather, flushing velocity guidelines should be developed according to various site-specific and system-specific conditions. This is discussed further in Chapter 6.

CHAPTER 6
CONCLUSIONS

INTRODUCTION

This chapter assimilates conclusions from the various project elements, including the Literature Review, CFD modeling, pilot testing, and field testing. These elements have offered several unique perspectives on various flushing phenomena, thereby helping to build a better overall understanding of the role of velocity in contaminant removal. Data gaps and recommendations for further research are also identified. Specific recommendations to utilities are provided in Chapter 7.

FLUSHING AS A MAINTENANCE TOOL

There are several different reasons why a utility may decide to implement flushing. The more common objectives of flushing are: (1) replacing stale water, (2) removing loose deposits, and (3) scouring (i.e. cleaning) the internal pipe surface. The focus of this study was on the latter two objectives.

The accumulation of both loose and adherent deposits can yield adverse water quality impacts. Loose deposits are susceptible to entrainment and suspension under normal hydraulic scenarios, such as flow reversals and velocity changes. These deposits can degrade the color and turbidity of finished water reaching the customers. Complex, adherent deposits may be stable during normal hydraulic conditions, but they may interact with bulk water in an undesirable fashion. For example, these deposits may exert disinfectant demand, produce disinfection byproducts, harbor microbes, and release taste-and-odor causing substances.

The behavior of loose deposits under various hydraulic scenarios was elucidated with CFD modeling and verified through pilot testing. Typical loose deposits of varying densities and size, including sand, GAC fines, and iron corrosion particles, were completely entrained from smooth pipes at velocities in the range 1.5-2.5 fps (0.5-0.8 m/s). For tuberculated mains, flushing effectiveness (measured as the fraction of pipe area cleaned) was reduced due to the presence of recesses and cavities. The CFD results predict significantly better particle removal than those of Stephenson (1989). For particles of specific gravity of 1.5 and 200 μm size in a 4-inch (100-mm) main, Stephenson (1989) recommends a minimum flushing velocity of 3.0 fps (0.9 m/s) for particle removal. In contrast, CFD modeling illustrated that flow velocities below 1.3 fps (0.4 m/s) would be sufficient for complete removal of these particles from smooth mains. For particles of specific gravity of 3.0 and 200 μm size in a 4-inch (100-mm) main, Stephenson (1989) recommends a minimum flushing velocity of 5.9 fps (1.8 m/s) for particle removal. However, CFD modeling illustrated that flow velocities of 2.0 fps (0.6 m/s) were sufficient to scour these particles from smooth surfaces. The CFD models developed around the smooth pipe agree well with the 'rule-of-thumb' value provided by Williams (2000). This rule-of-thumb, which is based upon previous pipe rig data, suggests that a mean velocity of 1.6 fps (0.5 m/s) is required to suspend most loose sediment. However, the CFD data suggests it should not be used for calculating flush rates for tuberculated pipes.

For the removal of more adherent deposits, pilot and field studies produced conflicting results with respect to the benefit of high velocity flushing. In a pilot-scale experiment where a pipe coupon containing extensive iron tubercles was subjected to flow velocities in the range 0-9 fps (0-2.7 m/s), there were no major sloughing events. This is likely due to the high hardness level of the source water and the calcium content of the tubercles in the pilot study, as described in Chapter 4. However, as illustrated in Table 6.1, significant additional amounts of material were removed during the field particle capture trials when the velocity was increased from 2-4 fps (0.6-1.2 m/s) to greater than 5 fps (1.6 m/s). The low velocity removal efficiency parameter used in Table 6.1 and defined in Equation 5-6 (Chapter 5) represents how much material could be removed at the low velocity relative to the total amount that could be removed by high-velocity unidirectional flushing. Assuming the loose material is removed at the low velocity (based on CFD modeling results), the additional material collected during the high velocity flushing may be more adherent and resistant to shear. While these data do not indicate that high-velocity unidirectional flushing was capable of completely cleaning the pipe surface, they indicate that some degree of scouring occurred.

Table 6.1
Benefit of high velocity in field particle capture trials

Utility-site	Pipe type	Pipe size	Low velocity removal efficiency
ANC-7	AC	6-in (150-mm)	26%
BEL-2	AC	6-in (150-mm)	44%
CLE-12	CLCI	8-in (200-mm)	29%
NNI-8	DI	8-in (200-mm)	24%
POR-5	UCI	8-in (200-mm)	0%

FACTORS AFFECTING FLUSHING PERFORMANCE

Flushing effectiveness is influenced by several factors, and the significance of each of these factors varies according to the utility-specific flushing objective. The theoretical and experimental approaches illustrated that the degree of pipe tuberculation and particle density are the two most critical factors for predicting the behavior of loose deposits during flushing. Of much less importance is particle size and pipe diameter. Table 6.2 provides a portion of the CFD modeling results and illustrates the relative impact of pipe condition and particle density.

Table 6.2
Velocity requirements for 90% pipe surface cleaning[1]

Pipe condition	SG = 1.1	SG = 3.0
Smooth	< 1.3 fps (0.4 m/s)	~ 2.6 fps (0.8 m/s)
Slight Tuberculation	< 1.3 fps (0.4 m/s)	~ 3.6 fps (1.1 m/s)
Heavy Tuberculation	~ 1.3 fps (0.4 m/s)	> 4.6 fps (1.4 m/s)

(1) Assumes 200-micron spherical deposits in a 4-inch (100-mm) main

Sediment characterization was performed on deposits collected from the field tests. The purpose of this task was to determine if different types of contaminants have significantly different hydraulic characteristics that would necessitate different flushing rates. Table 6.3 provides a summary of the characteristics of contaminants removed during the field trials.

Table 6.3
Average properties of system contaminants[1]

Parameter	Low velocity	High velocity
Specific gravity	2.90 ± 0.30	2.92 ± 0.08
Size	0.76 ± 0.28 mm	1.00 ± 0.27 mm
Iron fraction	40 ± 19%	39 ± 18%
Calcium fraction	2.4 ± 3.2%	1.1 ± 1.0%
Manganese fraction	0.41 ± 0.84%	0.62 ± 0.97%
Aluminum fraction	0.29 ± 0.28%	0.21 ± 0.18%

(1) Mean ± Standard Deviation

On average, deposits removed during the low velocity flush were coarse (0.5-1.0 mm) in size, while those removed during the ensuing high velocity flush were coarse to very coarse (1.0-2.0 mm) in size. Since 57-micron nets were used to capture the particles, the fine particle component is not included in the data set. The size ranges observed, especially those in the 1.0-2.0 mm size range, are too large to be introduced from the source water. It is more likely that these particles developed on the pipe wall, which would explain why it took a higher velocity and corresponding shear force to dislodge them.

Table 6.4 summarizes the specific gravities of deposits captured at low and (ensuing) high velocity flushes in several different water systems. In general, particle specific gravities fell within the range 2.5-3.5. From CFD modeling, it would be expected that if all deposits are loose, then those deposits removed during the high velocity flush would have a higher specific gravity than those removed during the preceding low velocity flush. However, the data do not support this trend. Rather, the data suggest that flushing performance is governed by particle characteristics other than density.

Table 6.4
Specific gravity of distribution system deposits

Site	Step	Pipe type	Average specific gravity	
			Low velocity (2-4 fps)	High velocity (≥5 fps)
ANC	7	AC	2.39	2.94
CLE	12	CLCI	3.10	2.81
TH	1	PVC	3.12	3.01
TH	2	UCI	2.94	2.87
TH	3	UCI	2.72	NA
TH	4	SI	3.15	2.96
Average			2.90	2.92

NA – Not Applicable

Deposits within the distribution system are constantly exposed to physical and chemical processes and forces. While in some instances these forces act to destabilize deposits and lift them into the water column, they can also act to immobilize deposits. Common examples of these latter processes include compaction, burial, and the formation of complex, cohesive deposits. These processes can occur independent of the type of deposit or contaminant.

If interparticle interactions are so significant, and ultimately mask the effect of particle specific gravity, it is natural to wonder what system conditions favor these processes. As discussed in the following section, the answer to this question is not trivial. A set of control variables was identified to help predict the need for a specific flushing velocity.

CONTROL VARIABLES

This study demonstrated that the factors that govern flushing effectiveness are intricately related to pipe type, water quality, level of treatment, routine hydraulics, and previous pipe cleaning practices. Carrière et al. (2002) also concluded that many site-specific factors contribute to flushing effectiveness. In their evaluation of loose deposits, the principal factors affecting deposit accumulations were the accumulation period, the pipe material, and the aggresivity of the water. The quantity and nature of the deposits varied greatly among the various networks and sampling campaigns. Williams (2002) also found that many factors affect sediment build-up in distribution mains including coagulant use, treated water quality, water hardness, water type and mains material.

Figure 6.1 illustrates the complexity associated with targeting specific flushing velocities. The top row of boxes represents those variables that can be identified and controlled by a water utility, and are hereafter referred to as "control variables". The middle row represents intermediate variables, which are considerably more difficult to evaluate and quantify. Flow arrows are used to define the presence of a significant relationship between sets of variables (boxes) in different rows. The following sections discuss the specific role of each of the control variables in Figure 6.1.

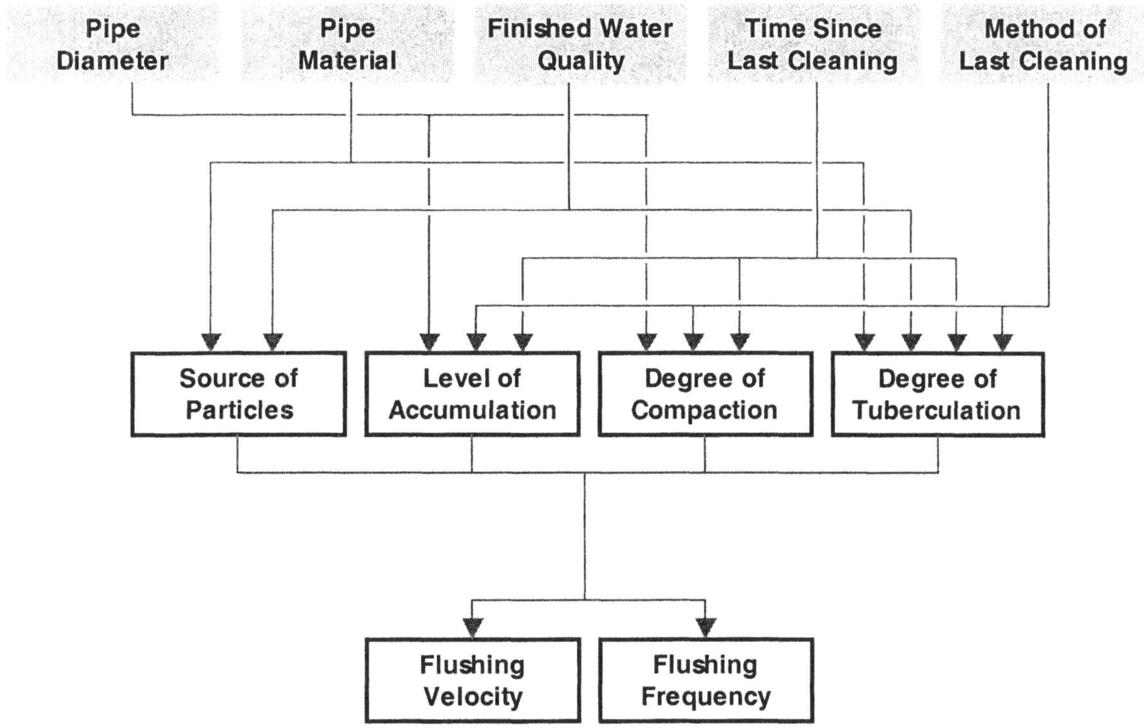

Figure 6.1 Variables impacting flushing effectiveness

Pipe Material

Pipe material is an important control variable since the piping itself can be a significant source of particles within the distribution system (and thus influence the rate of deposition), and the degree of tuberculation or pipe condition will impact the effectiveness of flushing at various velocities.

Source of Deposits

The field trials illustrate that water mains composed of unlined cast iron serve as a significant source of iron deposits. Particle capture trials performed on or near UCI mains consistently produced particles with the highest iron composition. Bulk water trials provided supporting evidence that the rate of accumulation of ferric deposits from UCI main corrosion is much higher compared to the rate of accumulation associated with treatment breakthrough or finished water iron contributions.

The Newport News field study (bulk water trials) provides insight into particle contribution and levels of accumulation over time. Hilton Village contains UCI mains roughly 80 years old, while Ivy Farms contains DI mains roughly 30 years old. Neither location has undergone (documented) flushing or cleaning. Further, both sites receive the same finished water. Because of age difference of these portions of the system, manganese levels were present in Hilton Village at levels 4.5 times greater than those in Ivy Farms. Iron, however, is present in Hilton Village at levels 91 times greater than those in Ivy Farms. If finished water contributions were the sole sources of iron, this difference would be expected to be closer to the 4.5 factor observed for manganese.

Degree of Tuberculation

Table 6.5 summarizes the average low velocity removal efficiencies of iron and manganese precipitates from four different pipe types. Low velocity removal efficiency provides an indication of the success of removing material by flushing at a low velocity (i.e., 2-4 fps [0.6-1.2 m/s]). According to CFD modeling predictions for *loose* deposits, the low velocity removal efficiency should be approximately 25-50% lower in tuberculated pipe compared to smooth or slightly tuberculated pipe. Field data indicate that the low velocity removal efficiency of iron deposits from UCI mains was approximately 44% lower than the efficiencies for all other combinations of pipe materials and deposits. The lower removal performance observed in the field is likely due to the fact that ferric deposits in AC and PVC mains are predominantly loose particulates originating from upstream UCI mains, and/or from finished water contributions. Conversely, for systems composed of UCI mains, iron deposits are contributed primarily by pipe corrosion and tubercles. The physical and chemical natures of deposits formed during the processes of corrosion and scale formation are considerably different than loose particle migration and settling, with the former being more adherent and difficult to remove at low velocities. With respect to manganese, particle occurrence is attributable solely to finished water contributions and subsequent oxidation, settling, and perhaps compaction. The removal efficiency within the tuberculated, unlined cast iron pipe was approximately 20% lower than removal within the smooth AC pipe, as shown in Table 6.5. However, manganese removal efficiencies were 23% greater than iron removal efficiencies within the unlined cast iron pipe. This difference in removal efficiencies suggests that both the degree of tuberculation and the degree of particle compaction/adherence impact flushing effectiveness.

Table 6.5
Average low velocity removal efficiencies (E_i) for bulk water trials

Pipe material	E_{Fe} (%)	E_{Mn} (%)
Asbestos cement	67.5%	65.0%
Polyvinyl chloride	66.5%	NT
Ductile iron	43.0%	39.0%
Unlined cast iron[1]	23.7%	46.7%

NT – Not Tested
(1) – Never been previously flushed

Finished Water Quality

Finished water quality characteristics will impact a system's flushing program in two ways: 1) finished water can be a significant source of distribution system particles, and 2) finished water quality can impact the degree of corrosion and tuberculation within distribution system piping.

Source of Deposits

Finished water quality may or may not be closely related to source water quality, depending on the type and extent of treatment provided. Common source water contaminants include sediment, sand, silt, organic matter, and metals. Under low flow conditions, these finished water contaminants may become deposits within the distribution system. Particle composition analyses during the field trials indicate that iron, manganese, calcium, aluminum, and magnesium are present in distribution system deposits.

Systems that purvey unfiltered surface water typically accumulate sediment at a much higher rate than comparable filtered water systems. Unfiltered ground waters may contain elevated iron and manganese levels. This suggests that certain systems may need to flush more frequently, but offers no implications about target flushing velocities. In fact, field study findings suggest that historical operational practices tend to dictate the appropriate flushing velocity more directly than treatment practices. This is discussed further in the sections Time Since Last Flush and Method of Last Flush.

Degree of Tuberculation

Finished water quality can also influence chemical reactions at the pipe-water interface. Some key water quality parameters in this regard are pH, temperature, alkalinity, and hardness. The corrosion of UCI pipes and deterioration of cementitious linings are two common examples of pipe surface reactions which can influence the types of deposits present and the effectiveness of flushing. A direct comparison between water corrosiveness and mass of material removed during flushing could not be made due to competing variables such as time since last flush, method of previous flushing, routine flow velocities, and other finished water quality conditions.

Time Since Last Flush (or Cleaning)

The length of time between flushing can have multiple impacts on flushing effectiveness. In general, longer time intervals mean more accumulated deposits and more complex, adherent deposits. If truly contaminant-specific velocities could be developed, a larger mass of accumulated deposits should be removed at the same flushing velocity as a smaller mass of the same deposits. This was not observed in the field due to numerous other variables which impact flushing effectiveness, such as compaction over time, particle interaction, and particle-pipe wall interaction. Field data indicate that time since last flush may be a controlling variable with regards to:

1) The level of deposit accumulation,
2) The degree of pipe tuberculation, and
3) The degree of particle compaction.

Level of Accumulation

The accumulation of conservative deposits, including sediment and metal precipitates, is positively related to the length of time between flushing (or cleanings), or time since installation if flushing has never been performed. During field-testing, hydrant discharge water consistently contained iron and manganese levels several orders of magnitude greater than the respective finished water concentrations, providing evidence of accumulation. Table 6.6 summarizes the range of iron and manganese particulates removed during field-testing, as well as the associated time intervals between cleanings.

Table 6.6
Normalized theoretical iron and manganese levels ($\hat{M}_i^*$) from bulk water trials

Parameter	Minimum $\hat{M}_i^*$	Time interval	Maximum $\hat{M}_i^*$	Time interval
Iron	3.8 mg/ft^2	6 years^1	2,176 mg/ft^2	80 years^2
Manganese	0.6 mg/ft^2	2 years^1	35.7 mg/ft^2	90 years^2

(1) Previously cleaned by unidirectional flushing
(2) No documented cleaning since installation

It is important to note that iron deposit accumulation may be controlled to a greater extent by pipe material versus finished water contributions, as described previously.

Degree of Tuberculation

In addition to the increased mass removed at Hilton Village, the low velocity flushing effectiveness was much lower at the Hilton Village site (11%) compared to the Ivy Farms site (43%) in Newport News, suggesting that a greater degree of tuberculation and/or particle compaction occurred over time. Again, this is consistent with CFD modeling predictions which indicate that removal efficiencies should be approximately 30% lower in tuberculated pipe (compared to smooth or slightly tuberculated pipe) in the velocity range of 2-4 fps (0.6-1.2 m/s).

Degree of Compaction

As deposits accumulate over time, they may change from loose particles to more cohesive or adherent deposits. This may explain why high velocity unidirectional flushing was highly beneficial for removing particulates from water distribution systems that had not been flushed or cleaned within 10 years, while high velocity flushing generally provided little to no additional benefit when conducted on systems that had been flushed within 4 years, particularly those that had been flushed by unidirectional methods. For the latter systems, the deposits were amenable to removal by low velocity flushing because they were generally still loose in nature.

Method of Last Flush (or Cleaning)

Methods for cleaning water mains include conventional flushing, unidirectional flushing, swabbing, and pigging. The aggressiveness of a particular cleaning method has a direct effect on the pipe surface topography, which in turn impacts particle removal performance. Heavily tuberculated pipe surfaces contain recesses and cavities that protect deposits from bulk flow shear forces. Consequently, the effectiveness of a flush becomes less sensitive to the flushing velocity as the degree of tuberculation increases.

Flushing is the least aggressive of the cleaning techniques, and is generally limited to mains with a nominal diameter less than or equal to 12 inches (300 mm). Field data provide some indication that water mains that have undergone high velocity unidirectional flushing within 6 years receive negligible benefit from additional high velocity flushing. Similar systems that had been flushed by conventional means within 6 years received significant benefit from high velocity flushing due to the ineffectiveness of particle removal associated with conventional flushing. Systems that routinely practice conventional flushing may still require initial high-velocity unidirectional flushing to remove accumulated deposits. Additionally, systems that routinely practice high-velocity unidirectional flushing may be using a great deal of resources for little improvement in distribution system water quality.

Pipe Diameter

Pipe diameter can have both direct and indirect impacts on flushing performance. The direct impact is due to the relationship between pipe diameter and flow hydrodynamics. The influence of pipe size and curvature on near-wall shear stress during flushing can be determined through the use of the following boundary shear equation (Oberoi 1994).

$$\tau = \frac{U^2 \rho}{(2.5 \cdot \ln(R/k) + 6.6)^2} \qquad (6\text{-}1)$$

Where:
τ = Boundary shear (Pa)
U = Bulk fluid velocity (m/s)
ρ = Fluid density (kg/m^3)
R = Hydraulic radius (m)
K = Hydraulic roughness (m)

Figure 6.2 illustrates the relationship between bulk fluid velocity and boundary shear. Data are provided for 4-inch (100-mm), 6-inch (150-mm), and 8-inch (200-mm) pipes. For this range of pipe sizes and flow velocities, boundary shear increases by roughly 10-20% for each 2-inch (50-mm) decrease in pipe size.

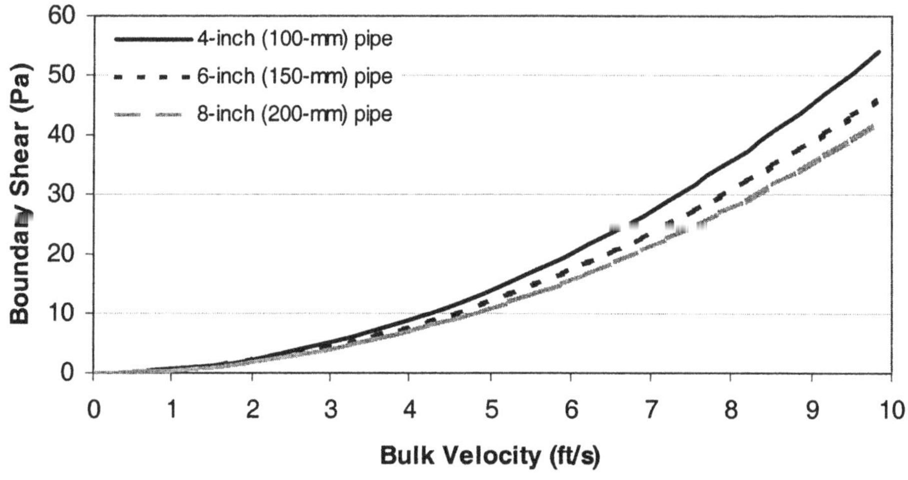

Figure 6.2 Effect of pipe diameter and bulk velocity on boundary shear (Adapted from Oberoi 1994)

Level of Accumulation and Degree of Compaction

Routine system velocities are dictated by the size of the mains and water demand requirements. Depending on the prevailing hydraulic conditions and flow velocities along a particular stretch of main, particles may settle and resuspend at differing rates. According to CFD predictions, particles with specific gravities of ≤ 3.0 should not settle at all in smooth pipes as long as the routine system velocities are always in the range of 2-3 fps.

STUDY LIMITATIONS AND DATA GAPS

The project's literature review showed that little published information was available on materials characterization. Most literature focused on adhered deposits due to their role in the passivation of the corrosion process. Research efforts to date have focused on describing the structure, chemical composition, and mechanisms of formation of pipe wall deposits. Relatively few studies address the adherence and size of pipe wall deposits. Of the few studies conducted on loose deposits in distribution systems, findings have focused on their characterization in terms of chemical composition and their occurrence in the distribution system. Little data are available on the physical characteristics of loose deposits, such as density and size. There is minimal published information addressing particles in the bulk water column. Examination of the nature of particles causing discolored water has been rare, perhaps due to difficulties analyzing what is often very small amounts of material or because the interpretation of the source of the particles is very difficult.

After development of several preliminary case studies as part of this Project, it became evident that the participating utilities had not selected flushing velocities based on site-specific conditions. The objective of the case studies was to capture information from utilities with long-term flushing programs and databases. Specific data of interest to this project included pre- and post-flushing water quality data, records of pipe ages/conditions, information on targeted particles, and records of flushing operations such as recorded pressures or flushing velocities. The case studies sought to answer specific questions such as whether the utilities used targeted velocities, how they determined acceptable velocities, how they judged the success of flushing, and whether they had correlated pipe velocities to particular materials.

CFD modeling was conducted as part of this Project to predict loose particle behavior only. While this effort produced extremely useful and experimentally-verified data, the nature of deposits in many distribution systems is very complex, and may exist as loose, cohesive, compacted, and/or adhered particles. Additionally, CFD modeling results are presented over the velocity range of 1.3 to 4.6 fps. The database should be expanded to analyzed velocities up to 8 fps. Additionally, CFD analyses should be conducted on a broader range of particle sizes, up to 2 mm.

The Project's pilot trials allowed for much better control over flushing variables, and the use of pipe sections allowed for visual inspection of pipe surfaces before and after each flush. Pilot testing was used to explore the effects of pipe-particle and inter-particle interactions. Experiments were effective for loose deposits on smooth pipe. However, experiments were not effective for cohesive deposits. The cohesive deposits tested actually behaved as loose particles. Also, the water used for testing was of considerably different quality than water for which the pipe coupons had been acclimated. The effects of these water quality differences on deposit properties are unknown.

The Project's field trials involved the most thorough characterization and quantification of flushed contaminants, but were also the most difficult to accurately execute. Relative to the coupons used during pilot testing, the flushing of water mains *in situ* required consideration of several additional variables. This fact, in combination with the invalidation of several data sets, made it difficult to draw sound conclusions about some variables and reduced the confidence in the trends that were identified. Another limitation of the field trials was the inability to see pipe sections before and after each flush. The low velocity removal efficiency results were useful for determining whether a higher velocity removed more material, but offered no information about the amount of material still within the system or the overall cleaning efficiency. The most meaningful field results were based on numerical integral approximations of bulk water data but in each case, only 2 or 3 data points were available.

RECOMMENDATIONS FOR FURTHER RESEARCH

This research study revealed the difficulties with putting science behind the art of flushing. Future research efforts should employ carefully engineered pilot studies. Pipe coupons from actual distribution systems should be utilized in order to accurately simulate real-world particles. Test water should also be representative of the system from which the pipe was exhumed. In order to ascertain whether flushing is an appropriate cleaning technique under specific conditions, mass balance studies should be conducted in conjunction with pilot tests. It is important to note that pilot study results will be most useful to utilities if they can match the conditions within their system to those applied in a specific study. Therefore, research should be conducted to further develop predictive capabilities of the independent variables previously discussed.

A decision tree should be developed to help utilities choose the appropriate flushing frequency based on system specific conditions. Criteria typically used to establish flushing frequency include water quality data; customer complaint records; a set time frame based on scientific literature, a consultant's recommendation or industry standards; water main material, age and condition; and/or flushing records such as time-to-clear and turbidity. The decision tree should also consider constraints to flushing such as water disposal issues, water conservation restrictions, pressure issues, service interruptions, and customer needs.

SUMMARY

Project findings suggest that it is not realistic to develop "contaminant-specific" flushing velocities. Rather, "site-specific" velocity recommendations may be developed since several processes appear to impact the stability/removability of deposits in distribution mains, and these may occur independent of the type of deposit. The findings from CFD modeling show that the most significant factors in determining the flushing velocity are the extent of pipe tuberculation and particle density. The pilot study trials have demonstrated that different types of contaminants have different degrees of adherence and cohesiveness, impacting the effectiveness of flushing as a control measure. The field trial results provide some indication of the influence of different variables on deposit accumulation rates and flushing performance. Perhaps the most important finding from the field trials is that "real-world" deposits can be loose, compacted, cohesive, or adherent, and these properties dictate the ability to scour and entrain specific particles. This study identified criteria and conditions that may be used to gauge the relative significance of these processes, and hence determine optimal site-specific flushing velocities.

Current industry flushing rates appear to be higher than required based on CFD modeling results. If the flushing objective is to remove loose deposits, then low velocity flushing (i.e., 2-4 fps [0.6-1.2 m/s]) is capable of removing most deposits from smooth and slightly tuberculated mains. With respect to tuberculated mains, higher flushing velocities produce a diminishing increase in particle removal performance. If the flushing objective is to scour and clean the pipe wall, high velocity flushing (i.e., ≥5 fps [1.6 m/s]) is generally needed. In the five (valid) particle capture trials conducted during field trials, use of the high velocity produced at least double the mass of discharged material. The relative benefit of using the high velocity can be gauged by considering pipe material, water quality, routine system hydraulics, and the time and method of previous flushing.

The following chapter provides a rationale for utilities to select site-specific flushing velocities based on knowledge of the system conditions and past practices (control variables).

CHAPTER 7
RECOMMENDATIONS TO UTILITIES

INTRODUCTION

The purpose of this chapter is to provide the reader with a simple decision tool for determining the most appropriate flushing practices specific to their system conditions. It is not intended to address emergency circumstances, such as contamination due to cross connections or regulatory water quality violations, but rather, it is geared toward the development of an optimized, routine flushing program. Ideally, a utility will have access to finished water and distribution system water quality data, distribution system mapping, historical O&M records, hydraulic modeling results, and other related resources.

Contrary to the initial aim of this research project, it is not realistic to establish universal contaminant-specific flushing velocities. While various distribution system particles may have distinct composition and density characteristics which can be assigned theoretical shear velocities required for scouring and transport, the reality of distribution system conditions requires consideration of particle cohesiveness, adherence to the pipe materials, burial and compaction, and site-specific hydraulics, as well as other variables. Thus, findings from this study have been used to develop more general flushing guidelines for use by water utilities. These guidelines are focused on two important topics:

(1) Selecting an appropriate flushing velocity
(2) Implementing an effective flushing protocol

Although boundary shear is the actual hydraulic parameter responsible for destabilizing deposits from the pipe surface, recommendations were generated based on velocity targets. While these two parameters are related (see Chapter 6), other variables such as effective pipe diameter and surface roughness also influence the shear force. However, these variables are not only difficult to assess, but they may vary along the length of a pipe or throughout the duration of a flush.

DECISION TREE TOOL

A flushing velocity decision tree tool, presented in Figure 7.1, was built around the five control variables previously discussed in Chapter 6 – pipe diameter, pipe material, finished water quality, time since last cleaning and method of last cleaning. The purpose of this section is to provide instruction on the use of the tool, guide the reader through the various pathways, and provide rationale and background information at each stage.

Flushing Objective

There are several different reasons why a utility may decide to implement flushing. This consideration is important because flushing velocity recommendations are specific to the objective. As such, this is the first question the user must answer. The three most common purposes of flushing are to: (1) replace stale water, (2) remove loose deposits susceptible to entrainment during routine hydraulic changes, and (3) scour and clean the pipe surface. Since water turnover (i.e., replacement of stale water) is not velocity-specific, this element is not incorporated in the decision tree.

Loose Deposit Removal

Loose deposits originate from both the finished water (sedimentation) and dissolution of pipe materials. These deposits are susceptible to entrainment under normal hydraulic scenarios, such as flow reversals and marked velocity changes. Failure to remove loose deposits can result in degradation of the color and turbidity of water at the customers tap.

Common examples of loose deposits include silt, sand, sediment, iron floc, iron sludge, iron precipitates, and manganese precipitates. Biofilm may also be associated with these particles. Based on the field tests conducted as part of this Project, the specific gravity of most of these deposits is in the range 2.5 to 3.5. Loose deposits are commonly found in unfiltered surface and groundwater systems, particularly in areas of low flow.

Velocity recommendations for removing loose deposits from distribution mains were extracted from CFD modeling results and are shown in Figure 7.1. The velocity ranges provided should be adequate to clean approximately 100% of the pipe surface for smooth and slightly tuberculated pipes, and at least 70% of the pipe surface for heavily tuberculated pipes. In order to optimize flushing velocities, information is needed on pipe condition. The following general definitions may be applied to the different pipe conditions:

- Smooth: 0% area reduction. Common examples of smooth pipe are plastic pipe, new AC pipe, newly lined DI pipe, and newly rehabilitated pipe.
- Slight Tuberculation: 0-1% area reduction. Common examples of slightly tuberculated pipe are older AC pipe in corrosive water, older DI pipe, and newer unlined CI.
- Heavy Tuberculation: 1-40% area reduction. Common examples of heavily tuberculated pipe are old, unlined CI and DI pipe and old, lined DI pipe in corrosive water.

According to Figure 7.1, lower velocities are recommended for removing loose deposits from mains. In fact, smooth pipes with routine velocities in the range of 2-3 fps (0.6-0.9 m/s) should be "self-cleaning"; in other words, particle deposition should not occur at routine velocities greater than 3 fps (0.9 fps). Slightly tuberculated pipes should be "self-cleaning" at routine velocities in the range of 3-4 fps (0.9-1.2 m/s). Routine velocities are defined here as pipe flow velocities that occur under average day conditions. The benefits of practicing low-velocity flushing include:

- Lower water consumption,
- Fewer water disposal issues,
- Lower impact on upstream residual pressures,
- Lower likelihood of creating subsequent red water problems due to scouring of scales and exposure of fresh pipe surfaces, and
- Lower likelihood of bursting mains in poor condition.

Pipe Wall Scouring

Often, utilities will perform flushing with the intent of scouring as much material as possible from the pipe wall. While complex, adherent deposits may be stable under normal hydraulic conditions, they may interact with bulk water in an undesirable fashion. Compacted and adherent deposits may incorporate silt, sand, sediment, corrosion products, tubercles, scale, and calcified deposits. Biofilm may also be associated with these particles. Factors that promote the formation of compacted and adherent deposits include low routine flow velocities, iron or cement-based pipes in corrosive water, poor finished water quality, and infrequent mains cleaning. These deposits may exert disinfectant demand, produce disinfection byproducts, harbor microbes, and release taste-and-odor causing substances. Also, the accumulation of these materials represents a loss in the carrying capacity of the pipeline.

Prior to implementing a pipe wall scouring flushing program, it is important for utilities to consider potential unwanted side effects that can be associated with aggressive flushing. While not the focus of this research effort, a summary of potential negative secondary impacts associated with pipe wall scouring is provided below. It may be necessary to flush a test portion of a distribution system if the utility is unsure of the likelihood for negative secondary impacts.

- Generation of red water – this is possible if tubercles are scoured and passivating conditions are not applied after flushing. The newly exposed surface can cause discolored water until the surface is repassivated. This can also result in increased disinfectant demand.
- Release of bacteria and opportunistic pathogens into water column – bacteria scoured from the pipe surface can be released into the water column. Unidirectional flushing is important to ensure that biofilms scoured during flushing are expelled from the system, and not merely transported to an adjacent pipe.
- Potential for transient low or negative pressures – under certain hydraulic conditions, high velocity flushing can cause the development of low or negative pressure waves that can travel throughout the distribution system. This increases the potential for intrusion or backflow if cross connections are present.

The remaining subsections are dedicated to utilities whose flushing objective is to scour and clean the interior pipe surface. Specific flushing velocities are recommended based on historical maintenance practices, routine hydraulics, and pipe condition.

Historical Maintenance Practices. Once a utility has decided to implement flushing to scour a particular section of main, they should give consideration to the previous flushing practices employed at that location. Four alternatives are provided: (1) never flushed; (2) flushed conventionally; (3) low velocity unidirectional flushing (i.e., 3-5 fps [0.9-1.5 m/s]); and (4) high-velocity unidirectional flushing (i.e., ≥5 fps [≥1.5 m/s]).

- Never Flushed - For mains that have never been flushed, there is a good probability that loose, compacted, and adherent deposits have developed over time. An initial round of high velocity flushing (≥5 fps [≥1.5 m/s]) is recommended to maximize deposit removal.
- Conventional Flushing - For systems that have conducted conventional flushing in the area, there may have been some benefit in terms of loose deposit removal. However,

since valve isolation was not employed, velocity targets may not have been attained and particles from nearby mains may have been transported throughout the area. There is a good probability that compacted and adherent deposits were not immobilized during previous flushing since high velocities are difficult to attain without valve isolation. An initial round of high velocity flushing (≥5 fps [≥1.5 m/s]) is recommended to maximize deposit removal from regions that were flushed using conventional approaches in the past.

- Low Velocity Unidirectional Flushing - For systems that have conducted low velocity unidirectional flushing, it is likely that most loose deposits and some compacted and adherent deposits were removed from smooth and slightly tuberculated mains. However, the field studies suggest that there is significant benefit to conducting one round of high velocity unidirectional flushing.
- High Velocity Unidirectional Flushing - For systems that have previously conducted high velocity unidirectional flushing (within the last 4-6 years), it is likely that all loose, compacted, and adhered deposits (that were amenable to removable by flushing) were removed. To determine the most appropriate velocity for subsequent flushing, consideration should be given to factors which influence sedimentation and compaction since the previous flushing event, including routine hydraulics, finished water quality, and pipe material.

Routine Hydraulics. Routine hydraulics and flow patterns have a significant impact on the processes of particle settling, compaction, and shearing. These factors should be given foremost attention when selecting a flushing velocity for mains that have previously undergone high velocity unidirectional flushing. Routine velocity can be used to predict the behavior of deposits introduced to the pipeline and generated since the previous flush. Three alternatives are provided: (1) 0-2 fps (0-0.6 m/s); (2) 2-4 fps (0.6-1.2 m/s); and (3) ≥4 fps (≥1.2 m/s). Hydraulic modeling may be required to evaluate routine velocities, based on average day demands.

- Low Routine Velocity (0-2 fps [0-0.6 m/s]) - According to CFD modeling, mains with routine flow velocities in the range 0-2 fps (0-0.6 m/s) are *not* self-cleaning. Therefore, particle deposition and compaction are possible, regardless of the pipe condition. The rate at which these processes will occur depends primarily on finished water quality and the elapsed time since the prior flush. Flushing velocity recommendations are geared towards pipe condition. For smooth and slightly tuberculated mains, unidirectional flushing should be conducted at a velocity of 4-6 fps (1.2-1.8 m/s). For heavily tuberculated mains, a flushing velocity greater than 5 fps (1.6 m/s) should be attained.
- Average Routine Velocity (2-4 fps [0.6-1.2 m/s]) - According to CFD modeling, flow velocities in the range 2-4 fps (0.6-1.2 m/s) are sufficient to prevent sedimentation on smooth and slightly tuberculated mains. Therefore, the extent of accumulation of loose and compacted deposits is expected to be low, and even high velocity flushing is unlikely to provide much additional benefit. With respect to tuberculated mains, deposition and compaction can still occur in recesses and cavities. Also, new adherent deposits such as corrosion tubercles may have formed on these mains. To optimize cleaning, a flushing velocity greater than 5 fps (1.5 m/s) is recommended.

- High Routine Velocity (≥4 fps [≥1.2 m/s]) - According to CFD modeling, mains with routine flow velocities above 4 fps (1.2 m/s) are self-cleaning. Particle deposition and compaction are very unlikely to occur. Therefore, the amount of removable material on the main is expected to be quite low, and flushing would not be expected to provide much additional benefit. Assuming that cleaning is still required, more aggressive techniques may be needed to clean these mains, such as pigging, rehabilitation, or replacement.

FUTURE FLUSHING

The Flushing Velocity Decision Tree (Figure 7.1) is set up for iteration. For the first use with a particular stretch of main, the user should begin at the top question (flushing objective). Once the user has reached a decision box, dashed lines are used to indicate the available flow pathways for making future flushing decisions.

SUMMARY

The initial aim of this research project was to develop contaminant-specific flushing velocities. Multiple approaches were used to determine contaminant-specific velocities including CFD modeling, pilot testing, and field testing. Research findings indicate that universal contaminant-specific flushing velocities are difficult to discern. While various distribution system particles may have distinct composition and density characteristics which can be assigned theoretical shear velocities required for scouring and transport, the reality of distribution system conditions requires consideration of particle cohesiveness, adherence to the pipe materials, burial and compaction, site-specific hydraulics, as well as other variables. Thus, findings from this study have been used to develop site-specific flushing velocity guidelines (versus contaminant-specific velocity guidelines) for use by water utilities.

A Flushing Velocity Decision Tree was developed to serve as a tool that allows systems to determine if they are currently flushing at appropriate velocities. The Decision Tree can be used by utilities that want to focus on prevention of customer complaints by removing loose deposits from the system (i.e., low velocity flushing), or by utilities that want to use flushing as a tool for cleaning distribution system piping (i.e., high velocity flushing). Potential negative secondary impacts associated with high velocity flushing were summarized as points of consideration prior to implementation.

Since most systems are comprised of various pipe materials that were installed over many years, there is no single flushing velocity that is optimal on a system-wide basis. Thus, utilities may want to tailor flushing programs to anticipated system conditions so that water quality objectives can be met while minimizing resources used.

Velocities ranging from 2 fps (0.6 m/s) to ≥5 fps (≥1.6 m/s) are recommended, depending on flushing objectives, pipe conditions, historical flushing practices, and routine hydraulics.

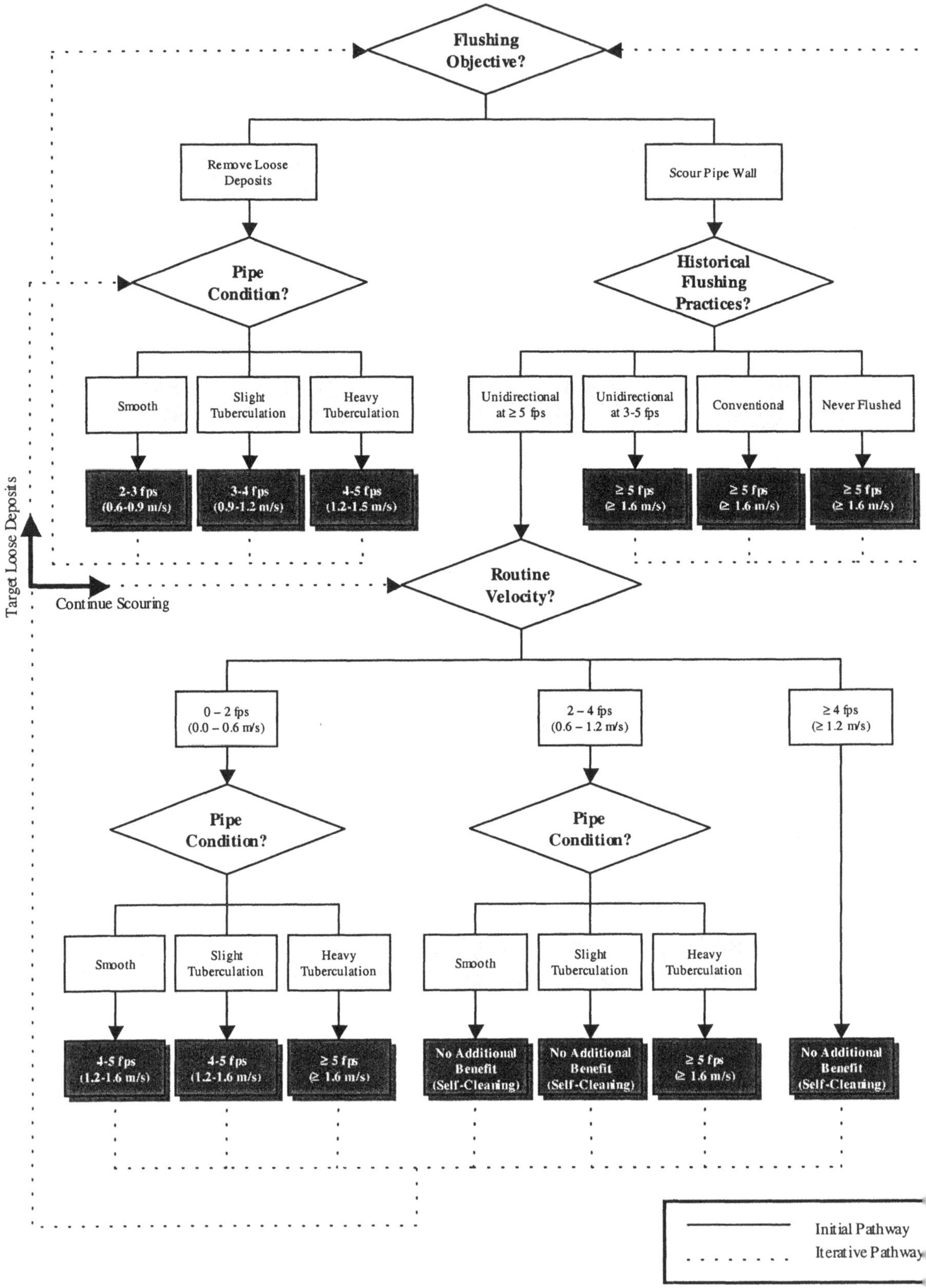

Figure 7.1 Flushing velocity decision tree (see accompanying notes and definitions for using decision tree tool on following page).

Figure 7.1 (cont.)

NOTES AND DEFINITIONS FOR USING DECISION TREE TOOL

General Note

This decision tree tool is to be used only in the development of an optimized, routine flushing program. It is not intended to address emergency circumstances, such as contamination due to cross connections or regulatory water quality violations.

Selecting Flushing Objective

- The three most common flushing objectives are to: (1) replace stale water, (2) remove loose deposits and (3) scour and clean the pipe surface. Since replacement of stale water is not velocity-specific, this objective is not incorporated in the decision tree.
- Removing loose deposits – this objective may be selected if mains were recently cleaned or rehabilitated (within 4 years) and the goal is to remove additional accumulated material to prevent customer complaints; or if a low velocity flush is needed to protect pipe lining material.
- Scouring pipe wall – this objective may be selected to remove complex, adherent deposits from pipe walls. Prior to implementing a pipe wall scouring flushing program, utilities should consider potential unwanted side effects that can be associated with aggressive flushing (i.e. generation of red water, release of bacteria and opportunistic pathogens into water column, potential for transient low or negative pressures that could allow pathogen intrusion or backflow if cross connections are present).

Estimating Pipe Condition

- Smooth: 0% area reduction. Common examples of smooth pipe are plastic pipe, new AC pipe, newly lined DI pipe, and newly rehabilitated pipe.
- Slight Tuberculation: 0-1% area reduction. Common examples of slightly tuberculated pipe are older AC pipe in corrosive water, older DI pipe, and newer unlined CI pipe.
- Heavy Tuberculation: 1-40% area reduction. Common examples of heavily tuberculated pipe are old, unlined CI and DI pipe and old, lined DI pipe in corrosive water.

Choosing Routine Velocity

Routine velocities are defined here as pipe flow velocities that occur under average day conditions. Hydraulic modeling may be required to evaluate routine velocities.

Flushing Duration

The utilities participating in this study flushed a minimum of one pipe volume during each step. KIWA and Dutch Water Utilities recommend flushing a minimum of three pipe volumes and until a (visual) inspection shows that the water is clear (Slaats 2001). If a field turbidimeter is used, a turbidity target of $\leq$ 5 NTU is recommended.

APPENDIX A
UTILITY SURVEY FORMS AND RESULTS

ESTABLISHING CONTAMINANT SPECIFIC FLUSHING VELOCITIES

AWWARF PROJECT NO. 2606

UTILITY SURVEY – TARGETED FLUSHING

March 2000

This survey is being solicited as part of the AWWARF Project, *Establishing Contaminant Specific Flushing Velocities*. The purpose of this survey is to collect information about utility flushing experiences that reflects specific observations and/or data collection with respect to targeted flushing velocities, or targeting specific pipe materials/deposit for removal. A certain amount of background and general information is also requested to categorize utilities by size, source water, and general flushing practices.

We appreciate your participation in this survey!

Please respond in the shaded areas only.

I. BACKGROUND INFORMATION

Utility Name:	
Location:	
Contact:	
Title:	
Phone Number:	
FAX Number:	
E-mail address:	
Service Population:	
Average daily production [mgd]:	
Primary or Consecutive Systems (Is your supply purchased?)	

Primary Source (please mark all that apply):

Surface Water:

	River		Lake		Reservoir
	Other (Please Describe)				

Groundwater:

	Well(s)		Spring		Infiltration Gallery
	Other (Please Describe)				

Source Quality (please write/type in average values or ranges):	
PH	
Alkalinity (as $CaCO_3$)	
Calcium Hardness (as $CaCO_3$)	
Iron	
Manganese	
Ca	
Mg	
Average Turbidity	
Total Color (CU)	
Total Organic Carbon	
Specific Ultraviolet Absorbance SUVA (if known)	

Treatment Used (please provide a brief description, including chemicals used):

Filtration (please write/type in answers):	
Type of filter(s)	
Filter rate	
Coagulant type and dose	
Filter effluent measurements:	
Turbidity (range)	
Particle counts (range)	
Primary disinfection (please write/type information in space provided):	
Disinfectant used:	
Typical dosages:	
Typical CT achieved (Winter vs. summer if appropriate)	
Corrosion Control Treatment (please write/type information in space provided):	
Type of treatment:	
Typical dose:	

90th percentile lead and copper concentrations from last compliance sampling:			
	Pb		Cu

Residual disinfectant (please write/type information in space provided):	
Type of disinfectant:	
Typical dosage:	
Target chlorine:ammonia ratio (if chloramines):	
Range of residuals in distribution systems (winter versus summer)	
Finished Water Quality (please write/type information in space provided):	
PH	
Alkalinity (as $CaCO_3$)	
Total Hardness (as $CaCO_3$)	
Iron	
Manganese	
Average Turbidity	
TOC	
Ca	
Mg	
Particle count (and size range)	

II. DISTRIBUTION SYSTEM INFORMATION

Total miles of distribution main	

For each type of distribution main listed below, indicate the range of sizes (pipe diameter) and the % of total miles of main (please write/type information in space provided):

Type	Size (in)	Approximate % of total miles
Cast iron-lined		
Cast iron-unlined		
Ductile iron-lined		
Ductile iron-unlined		
Steel		
PVC		
Asbestos/Cement		
Concrete		
Other (describe)		

Distribution System Isolation Valves (please write/type information in space provided):	
Is an inventory kept for isolation valves? (yes/no/unknown)	
Last update? (year)	
Total # valves, size range (in.)	
Is there a numbering scheme?	
Is there a valve exercise program?	
Frequency of valve exercising?	
Hydrants (please write/type information in space provided):	
Is an inventory kept for hydrants? (yes/no/unknown)	
Last update? (year)	
Total # active hydrants	
Is there a numbering scheme?	
Is there a hydrant maintenance program?	
Frequency of maintenance (% total hydrants per year)	

Can you identify the source of sediment/contaminant material found in the distribution system?						
Is it due to lack of adequate surface water treatment?	Yes		No		Unknown	
If yes, please explain briefly in the space provided below:						
Is it due to filter breakthrough?	Yes		No		Unknown	
Is it due to alum(or other coagulant) precipitation?	Yes		No		Unknown	
Is it due to direct intrusion into the distribution system from outside sources?	Yes		No		Unknown	
Other? (Please Explain)						

III. FLUSHING INFORMATION

Does your system have a regularly scheduled flushing program?	Yes		No	
If not, do you flush on an irregular basis?	Yes		No	
How do you select the area(s) to be flushed? (Please write/type information in space provided):				

Once the area has been chosen, do you manipulate valves to direct/limit the flows (i.e. unidirectional flushing)? (Please write/type information in space provided):						
Are there specific flushing criteria such as:						
Flow rates at hydrant	Yes		No		Unknown	
Minimum pressure	Yes		No		Unknown	
Targeted pipe velocities	Yes		No		Unknown	
Time	Yes		No		Unknown	
Other (please explain)						
Are there flushing objectives, such as:						
Restore disinfectant residual	Yes		No		Unknown	
Reduce disinfection by-product precursors	Yes		No		Unknown	
Reduce number of bacteria	Yes		No		Unknown	
Reduce turbidity	Yes		No		Unknown	
Eliminate discolored water	Yes		No		Unknown	
Eliminate taste and odors	Yes		No		Unknown	
Other (please explain)						
Who does the flushing?:						
Normal maintenance personnel	Yes		No		Unknown	
Special flushing crews	Yes		No		Unknown	
Outside contractors	Yes		No		Unknown	
Other (please explain)						
Are there written procedures for the program?	Yes		No		Unknown	
How were these procedures developed (please explain)						
Do the flushing crews carry these written procedures into the field?	Yes		No		Unknown	
Have the flushing crews received training on flushing procedures?	Yes		No		Unknown	
Is flushing typically conducted during the day?	Yes		No		Unknown	
Are all parts of the system flushed using the same protocols?	Yes		No		Unknown	

What triggers the decision to flush? (Please mark all applicable items)	
Schedule (annual, semi-annual, etc.)	
Water quality complaints	
Water quality sampling results	
Water main installation or repair	
Other (please explain)	

What information is collected during the flushing? (Please mark all that apply)	
Flow	
Pressure	
Time to clear	
Water quality information	
Other (please explain)	

Do you collect any other information (for example, chlorine residual, iron or manganese, sediment, corrosion control agent, etc.)? (Please mark each answer)						
Before flushing	Yes		No		Unknown	
During flushing	Yes		No		Unknown	
After flushing	Yes		No		Unknown	

Please list the types of data/information collected in the space provided below:

Is hydrant flush water dechlorinated before disposal?	Yes		No		Unknown	
If yes, do dechlorination methods restrict flows or velocities?	Yes		No		Unknown	

What triggers the decision to stop flushing? (Please mark each answer)	
Time	
Clarity (turbidity; suspended solids; visual)	
Disinfectant residual	
Other (please explain)	

Is the effectiveness of the flushing evaluated?	Yes		No	
If yes, how is it evaluated?				
Is this documented?				
Do you have a special program for dead-ends?	Yes		No	
Do you have a special program for continuously open blowoffs?	Yes		No	
Do you have a special program for periodically (scheduled) opened blowoffs?	Yes		No	
Do you have a special program for spot flushing?	Yes		No	

Within the past 5 years, how many times have you: (please write/type response)	
Completely flushed the entire distribution system?	
Flushed entire regions or zones?	
Spot flushed localized areas?	

How is your flushing program affected by water conservation efforts, droughts, water use restrictions, etc.? (Please write/type response)

Are there any special problems or problem areas associated with the flushing?	Yes		No		Unknown	
If yes, please explain:						
Do customers have specific concerns about flushing?	Yes		No		Unknown	
If yes, please explain:						
Have you tried any alternative flushing procedures (such as air or a combination of air/water)?	Yes		No		Unknown	
What were they?						
When and how often were they tried?						
How was the effectiveness or lack of effectiveness measured?						

IV. Targeted Flushing Velocities

Have specific velocities been established for certain water quality objectives?	Yes		No	
If yes, please list and describe				

How were velocities established? (Please mark appropriate answers)

Trial and error		Field Testing		Other	

Are velocities varied based on: (Please mark all answers that apply)

Pipe type		Pipe material		Pipe age and condition	

Please describe any special studies that have been done to determine appropriate velocities

Is a protocol available for any special studies that have been done?	Yes		No		Unknown	
During any special studies, did you collect sediment/contaminant material?	Yes		No		Unknown	

During any special studies, did you analyze or characterize collected sediment/material?	Yes		No		Unknown	
Do you have protocols for characterization?	Yes		No		Unknown	
Are there parts of your distribution system where adequate pressure and/or velocities are unattainable?	Yes		No		Unknown	
If yes, have you used alternative methods of cleaning the pipe?	Yes		No		Unknown	
What were they? (Please explain below)						
Thank you for completing this survey.						
Would you like to receive a copy of the survey results?			Yes		No	
In order to track your participation as "in-kind" services to AWWARF, please add the total hours spent by all staff to complete this survey and provide it in the space below.						
Man hours to complete survey						
Date						

UK Survey Form
Mains flushing best practice –questionnaire to the industry

Background

Below are listed a series of questions asked of water industry personnel involved in mains flushing programmes in order to gather some initial background information about current flushing practices within the UK water industry.

Questions

1. Do you perform systematic mains flushing? If N, go to question 8, if Y, question 2

2. For what reason?

3. How do you decide it is required?

4. How do you decide the return frequency?

5. Describe in detail the planning process and the operational method.
   Eg
   - Do you plan flushing with the aid of a network model?
   - Do you use all available hydrants or a selection?
   - How do you select the hydrants?
   - Do you aim to achieve a specific water velocity and if so, what velocity?
   - Do you flush for a set time period?
   - What diameter ranges do you flush?
   - What mains materials?
   - Do you follow written company guidelines?
   - Are there any constraints applied such as: work to be performed at night time only?
   - How do you warn customers?
   - Have you got any automatic hydrant flushing equipment?

6. Do you perform any monitoring, either of water quality or customer perception, to establish the need or assess the success of the flushing? What criteria do you use? If N, go to question 8, if Y, question 7

7. If yes, what monitoring, with what outcome?

8. Do you perform regular flushing at specific locations? If N, go to question 16, if Y, go to question 9.

9. For what reason?

10. How do you decide on location?

11. How do you decide on frequency?

12. How do you decide when to stop regular flushing at a particular location?

13. Describe in detail the planning process and the operational method.

    Eg

    - How do you select the hydrants?
    - Do you aim to achieve a specific water velocity and if so, what velocity?
    - Do you flush for a set time period?
    - What diameter ranges do you flush?
    - What mains materials?
    - Do you follow written company guidelines?
    - Are there any constraints applied such as: work to be performed at night time only?
    - How do you warn customers?
    - Have you got any automatic hydrant flushing equipment?

14. Do you perform any monitoring, either of water quality or customer perception, to establish the need or assess the success of the flushing?

15. If yes, what monitoring, with what outcome?

16. Do you do any other mains cleaning, either systematic or of hotspots?

17. Describe what you do, how you do it and how you decide it needs doing. How do you decide on mains cleaning technique?

18. Have you costed the processes you use? What are the costs and what are the various elements?

19. Do you use contractors or direct labour? If a mixture, which do you use for what?

20. Are there any particular operational constraints, eg night time, disposal of waste water? How do you deal with them?

21. Any special kit in use or novel methods? If so, what?

22. Any further details which you feel would be worthy of note in a "best practice" flushing document?

UTILITY SURVEY RESULTS

The purpose of the project survey was to collect information about utility flushing experiences that reflect specific observations and/or data collection with respect to targeted flushing velocities, or targeting specific pipe materials/deposits for removal. A certain amount of background and general information is also discussed to categorize utilities by size, source water, and general flushing practices. Two survey forms were developed, one for utilities in the United States and Canada (North American survey), and one for United Kingdom utilities (UK survey).

A total of 36 surveys were sent to U.S. and Canadian utilities. Twenty-three (64%) were completed and returned. Five of the North American respondents were classified as medium size utilities serving less than 50,000 connections and the remaining eighteen were classified large utilities serving greater than 50,000 connections. The responding utilities were located in 10 U.S. states, and the province of British Columbia, Canada. Fifteen of the responding utilities reported surface water as the sole source of supply, 1 reported ground water as its sole source and the remaining seven reported their source water as a combination of surface and groundwater. Four of the reporting utilities purchase 100% of their water through wholesale agreements with regional suppliers; sixteen utilities owned their source of supply; and three reported a combination of direct supply and wholesale purchase to varying degrees.

A utility flushing survey was already being carried out in UK as part of another research project. These data were reported here rather than repeating the exercise. The only difference between the UK and U.S. surveys was that UK surveys did not request background information on distribution system characteristics, sediment sources and maintenance practices. The survey requested specific information in fourteen different areas.

- Targeted Flushing Velocities
- Distribution System Physical Characteristics
- Distribution System Inventory and Maintenance Practices
- Sources of Sediment and Contaminants
- Formalized Flushing Programs
- Flushing Objectives
- Criteria for Selection of Flushing Areas
- Triggers To Implementation of Flushing Activity
- Planning Techniques
- Flushing Methods
- Alternatives to Flushing
- Flushing Frequency
- Flushing Constraints
- Data Collection
- Evaluation of Flushing Program

Distribution System Physical Characteristics

The North American reporting utilities represent more than 30,000 miles (48,000 km) of water distribution mains, ranging in diameter size from 1 inch (25 mm) to 108 inches (2.7 m).

Seventeen of 23 utilities provided information on miles of main by type. Table A.1 is a compilation and summary of date provided by the seven reporting utilities. This information was not requested in the UK survey.

Table A.1
Summary of distribution system materials compiled from seven reporting utilities

Material	Miles	Fraction of total
Cast iron, unlined	4,912 +	20.2%
Cast iron, lined	6,840 +	28.2%
Ductile iron, lined	3,201 +	13.2%
Ductile iron, unlined	3,000 +	12.4%
Combined unlined cast iron and ductile iron	3,125	12.9%
Asbestos cement	1,008	4.2%
Polyvinyl chloride	323	1.3%
Steel	586	2.4%
Concrete	1,040	4.2%
Other	236	1%

Twenty of 23 utilities provided data on the number and size of existing valves for a total of approximately 421,000 isolation valves ranging in size from 0.75-inch (19-mm) to 72-inch (1.8 m).

Distribution System Inventory and Maintenance Practices

Distribution system isolation valves are inventoried by 19 of 23 utilities in the North American survey. Ten of these 19 utilities update their inventory records continuously and an additional 6 utilities have updated their records as recently as 1999. Twelve of 23 utilities have a numbering scheme for isolation valves. Sixteen of 23 utilities have a valve exercise program. Valve maintenance statistics are summarized in Figure A.1. Several utilities exercise critical valves on an annual basis and non-critical valves every 3 years. This information was not requested in the UK survey.

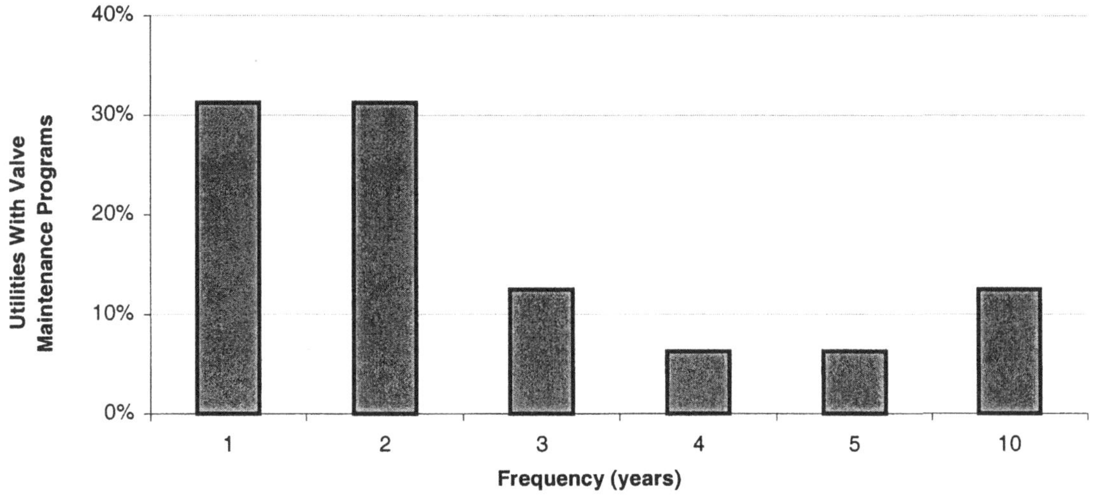

Figure A.1 Frequency of valve exercising

Twenty-one of 23 utilities in the North American survey keep an inventory of active hydrants. A total of 181,000 hydrants are represented in this survey. Just over 50% of utilities update their inventory records continuously (12 of 21 utilities) and others report the last update was completed in 1999 (9 of 21 utilities). Eighteen of 23 utilities use a numbering scheme for hydrants. Twenty of 23 utilities have a hydrant maintenance program. The most common maintenance frequency reported was 100% of hydrants maintained per year (7 of 23 utilities with hydrant maintenance programs). Hydrant maintenance statistics are summarized in Figure A.2. This information was not requested in the UK survey.

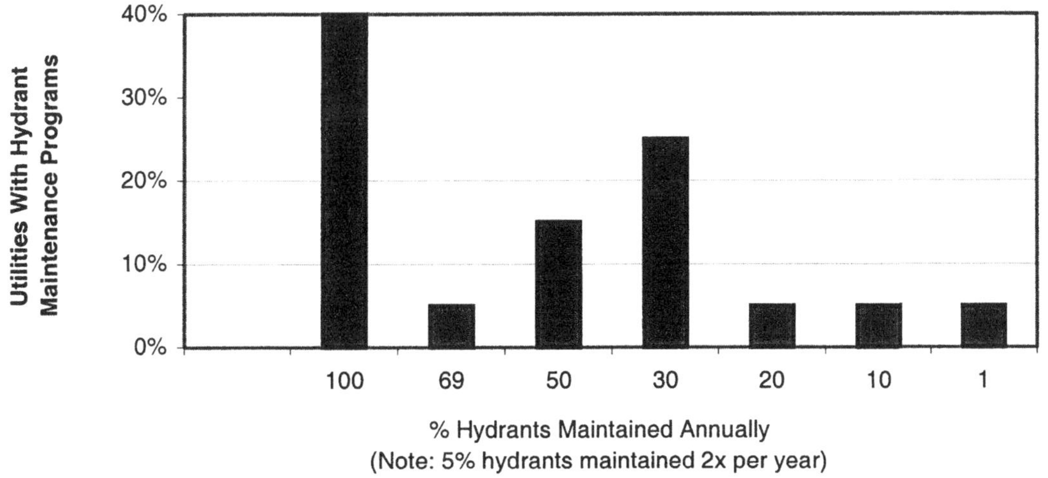

Figure A.2 Hydrant maintenance statistics

Sources of Sediment and Contaminants

Regarding sources of sediment and other contaminants in the distribution system, a lack of surface water treatment is the most frequent source reported in the North American survey (7 of 23 utilities). Other sources of sediment and contaminants are listed below.

- Precipitation of dissolved iron and manganese by chlorine residual
- Corrosion of distribution pipes and fittings
- Coagulant precipitation
- Pipe deterioration/aging
- Construction debris
- Corrosion of cement lining
- Low velocity
- Calcium carbonate precipitation
- Tubercles breaking apart

Three of 23 utilities surveyed in the North American survey collected sediment samples during flushing and 2 of these 3 utilities analyzed the sediment samples. Only 1 utility has a protocol for sediment characterization. This information was not requested in the UK survey.

Formalized Flushing Programs

Most utilities included in the North American survey have regularly scheduled flushing programs (20 of 23 utilities). The remaining 3 utilities flush on an irregular, as needed basis. In the UK survey, 9 of 15 utilities conduct systematic main flushing. Flushing activity in the UK was reported as increasing in response to UK Drinking Water Regulation's requests for operation and maintenance policies and practices.

Flushing Objectives

Flushing objectives identified in the surveys are listed below in order of the frequency cited:

- Eliminate discolored water
- Restore disinfectant residual
- Reduce turbidity
- Eliminate tastes and odors
- Reduce the number of bacteria
- Reduce DBP precursors
- Remove sediment
- Comply with regulations
- Maintain water quality
- Decrease chlorine demand
- Respond to customer complaints
- Reduce corrosion inhibitor build-up

- Eliminate stale water
- Response to animal activity
- Eliminate lime deposits

Criteria for Selecting Flushing Areas

Responding utilities reported the following criteria in prioritizing areas of the distribution system to be flushed. Figure A.3 shows a comparison of the different criteria and the frequency in which they are used. This data was reported only by North American utilities:

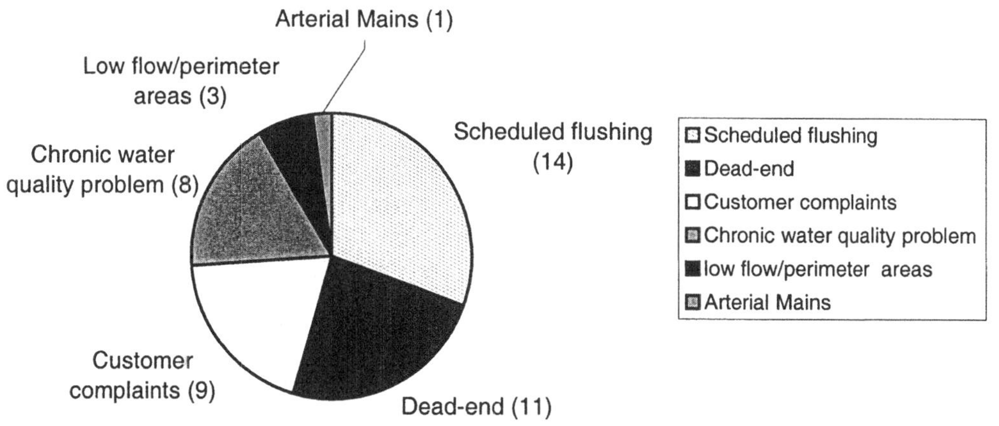

Total number exceeds number of respondents because some utilities reported more than a single criteria.

Figure A.3 North American utility reported criteria for flushing area selection

Triggers to Implement Flushing Activities

Responding utilities reported specific occurrences that triggered flushing activity. These triggers are listed below according to the frequency of being reported by the utilities:

North American utilities	UK utilities
Customer complaints on water quality	Customer complaints on water quality
Water quality sampling results	Water quality sampling results
Water main installations/repairs	Known problem areas
Schedule	Inspection
Cross connection incident	Operational activity
Other triggers (taste and odor, low turnover, dead ends, sediment/debris, manganese deposits, excessive iron levels)	Other triggers (taste and odor, low turnover, dead ends, sediment/debris, manganese deposits, excessive iron levels)

Thirteen of 15 UK utilities have a regular flushing program at specific locations in the distribution system. Flushing is most often performed to address discolored or dirty water (9 of 13 UK utilities) or customer complaints (5 of 13 utilities). The flushing locations and frequency are most often determined based on customer complaints and water quality monitoring results. Flushing is typically continued at these locations unless the main is rehabilitated or the operational or design changes take place (i.e., eliminate dead-end via looping). One utility has automatic flushing equipment in one part of their system. The decision to stop flushing is most often triggered by clarity of the water (22 of 23 utilities surveyed). Other parameters used to stop flushing include, in order of number of responses:

- Disinfectant residual
- Preset flushing time
- Property damage
- Inability to dispose of flush water
- Phosphate level
- Temperature
- Taste and odor

Planning Techniques

Eighteen of 23 North American utilities reported having written flushing procedures. Of those with written procedures different methods of development were reported including in-house staff, special teams, consultants, research, and trial and error. Fifteen of 18 utilities with protocols used the same protocols to flush all parts of the system. For those utilities reporting a written protocol, the following criteria were included in the procedures listed by frequency of inclusion in protocols:

- Minimum pressure
- Targeted pipe velocities
- Flow rates at hydrants
- Flushing time
- Clarity of water
- Increased chlorine residual
- Turbidity < 1.5 NTU
- Pipe volume turnover
- Lower temperature

For 22 of 23 utilities, flushing crews have received special training on flushing procedures. Flushing crews at 11 of 23 utilities used written procedures in the field. For 18 of 23 utilities, flushing is conducted during daylight hours.

Eighty-three percent (19 of 23) of North American utilities surveyed conducted flushing activities with regular maintenance personnel; the remaining 4 utilities used specially trained and assigned crews. In the UK, direct labor is commonly used for flushing at 13 of 15 reporting utilities; air scouring (7 of 9 utilities that use air scouring); and swabbing (5 of 6 utilities that use swabbing). The use of outside contractors to perform flushing activities was reported by 2 of 23

North American utilities, but only for flushing of newly constructed mains. One of 15 UK utilities used contractors. UK respondents also reported that contractors are occasionally used to conduct air scouring (1 of 9 utilities), pigging (1 of 2 utilities) and swabbing (2 of 6 utilities).

Most of the UK utilities use written company guidelines for flushing procedures. Of 9 UK utilities that conduct systematic mains flushing, 3 utilities plan flushing with the aid of a network model. Many water companies are starting to analyze customer complaint records and water quality data to establish priority lists for planned flushing. One of the 9 utilities has automatic flushing equipment in one part of their system. Four of these 9 utilities use all hydrants and 5 utilities use selected hydrants. Hydrants are selected in various ways including GIS, in-house planning, ease of access, end hydrants, and those that would minimize disruptions. Four utilities flush mains of all materials, 4 utilities flush predominantly iron mains, and 1 utility flushes only PVC and AC mains. The range of pipe diameters flushed varies widely, from 4-inch (100-mm) diameter and less, to all pipe diameters. Customers are typically notified of flushing activities by leaflet or letter (6 of 9 UK utilities). Six UK utilities address customer-driven constraints by flushing at night.

Flushing Methods

Unidirectional flushing methods are used by 13 of the 23 US utilities surveyed. Five additional utilities have used unidirectional flushing to a limited extent, either for pilot programs or for selected circumstances. Special flushing programs are commonly adopted for spot flushing (16 of 23 utilities), dead ends (14 of 23 utilities) and periodically opened blow-offs (14 of 23 utilities). Only 4 of 23 utilities have a special flushing program for continuously open blow-off ports.

Alternatives to Flushing

In the survey of UK utilities, utilities reported using the following alternatives to flushing:

- Air scouring (60%)
- Swabbing (40%)
- Pigging (13%)

Five utilities have developed cost estimates of these alternative methods. Innovative flushing techniques used in conjunction with the alternatives include a WRc patented Flow Jet method, a GIS-generated flushing program, and use of hydrant nets or baffle boxes to filter highly discolored water. None of the 23 North American utilities surveyed have tried alternative flushing methods such as air or an air/water combination.

Flushing Frequency

Flushing frequencies reported in the US survey are summarized below.

- Completely flush whole system 1 or 2 times per year (7 of 23 utilities, 30.4% utilities surveyed)

- Completely flush whole system 3 or more times per year (6 of 23 utilities, 26.1% utilities surveyed)
- Flush entire zones of distribution system 1 or 2 times per year (8 of 23 utilities, 34.8% utilities surveyed)
- Flush entire zones of distribution system 3 or more times per year (10 of 23 utilities, 43.5% utilities surveyed)
- Localized spot flushing 1 or 2 times per year (1 of 23 utilities, 4.3% utilities surveyed)
- Localized spot flushing as needed (5 of 23 utilities, 21.7% utilities surveyed)
- Localized spot flushing numerous times (13 of 23 utilities, 56.5% utilities surveyed)

Flushing frequency reported in the UK survey was sporadic and based not on a set time frame but on the following criteria:

- Water quality sampling results
- Turbidity trends
- Inspectors
- Customer complaints
- Trial and error
- Annual assessment of high risk areas

Flushing Constraints

Fifteen of 23 North American utilities reported that no legal restrictions were in place that limited flushing activity due to conservation efforts, droughts or other issues. Some utilities did, however, use self-imposed restrictions. Five utilities do not flush during drought conditions except for the possible exception of necessary spot flushing. Three utilities reported that flushing was avoided from May to September to avoid summer conservation restrictions and/or higher demand periods. And one utility reported that it ceases flushing when raining due to a combined sewer overflow issue.

Only 8 of 23 utilities reported being required to dechlorinate flushed water before disposal. Of these utilities, only 1 utility reported experiencing restricted flows or velocities due to the dechlorination process.

Survey respondents also listed the following special issues that limit or restrict flushing activity:

- Areas where no sewers or storm drains are available
- Water disposal considerations
- Flooding/property damage
- Big mains/low pressure
- Disruptions in downtown area
- Sensitive customers i.e., hospital
- Notifying customers prior to flushing

Data Collection

Eighteen of 23 North American utilities reported collecting water quality and operational data before flushing; thirteen collected data during flushing; and sixteen collected data after flushing. Water quality parameters reported as collected include:

- Time to clear
- Flow rate
- Chlorine residual
- Pressure
- Operations data (main size, length, which valves operated)
- Turbidity
- Color
- pH
- Bacteria
- Phosphate level

Five of 9 UK utilities conduct water quality monitoring before or after flushing. Various parameters are measured, including turbidity and chemical analyses. One utility uses hydrant nets to capture flushed debris and characterizes this debris by type of organism and sediment type. Some of the UK utilities use water quality data to evaluate regulatory compliance and the need to reflush.

Evaluation of Flushing Program

Five UK utilities evaluate flushing by conducting customer surveys. The number and location of complaints are recorded and reviewed on a periodic basis. In North America, the effectiveness of flushing is evaluated by 17 of 23 utilities surveyed. This evaluation is based on the following information:

- Water quality data
- Customer complaint data
- Historical data

Customer concerns about flushing are also tracked by 17 of the 23 North American utilities. The concerns reported in the survey include:

- Discolored water complaints
- Wasting water/need for flushing
- Impact on water quality during flushing
- Loss of pressure
- Property flooding
- Environmental concerns
- Glare of lights from flushing operation
- Inconvenience
- Lack of notification

APPENDIX B
CFD MODELING METHODOLOGY

APPENDIX B
CFD MODELING METHODOLOGY

1. METHOD

1.1. OVERVIEW

The work comprised the following stages:

1. Build and solve three CFD models (smooth, slightly tuberculated and tuberculated) for each of three pipe sizes: 4", 6" and 8".
2. Develop a methodology to calculate the lifting velocities for various particles in the above pipes.
3. Derive K to enable calculation of lift, by calculating the Saffman (1) lift, using pipe test rig trial data. Details of the pipe rig trials are described in the third periodic report.
4. Use K to calculate the fractional pipe area cleaned, for the three pipe conditions, for each of the three pipe diameters, for particles of different sizes and densities.
5. Present results graphically and provide raw data as a database.

1.2. BUILDING CFD MODELS

The CFD technique involved the following stages:

- Generate the computational geometries and grids
- Calculate the inlet velocity flow profiles
- Calculate the steady state-flow patterns
- Calculate and export the local velocity and velocity gradient values

Each pipe comprised between 59,000 and 79,000 unstructured tetrahedral cells. The flow solutions were converged to give residuals below 10^{-4} SI units.

1.2.1. Geometry

For each pipe diameter, three pipe sections were built. These represented a new, smooth pipe, a slightly tuberculated pipe and an old, highly tuberculated pipe. These are shown in **Figure B.1**, below.

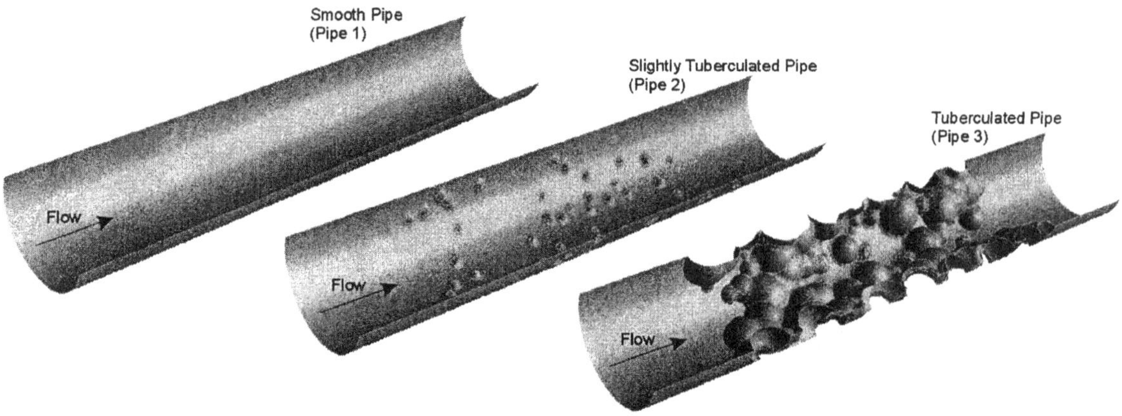

Figure B.1 Cutaway pipes showing the three tuberculation states studied

The tubercles were created as spheres with a range of random diameters whose centres lay on the pipe surface. The spheres were placed with a fixed axial spacing and distributed randomly around the circumference of the pipe.

The diameter ranges and axial distances are given in B.1, below, for the 4" pipe. For the larger 6" and 8" pipes, the 4" model mesh was scaled up.

Table B.1
Pipe configurations

Pipe no.	Tubercle radii (mm)	Axial spacing (mm)	Total number	% Area reduction
1	n/a	n/a	0	0
2	3-5	3	100	0.6
3	3-20	1	300	36

Each of the pipes was five pipe diameters in length. The first and last pipe diameters were left without tuberculation to allow a reasonably smooth flow through the inlet and outlet. The pipe dimensions are summarised in **Table B.2**, below.

Table B.2
Pipe dimensions

Pipe no.	Pipe diameter (mm)	Pipe length (m)	Tuberculated length (m)
1	100	0.5	0.3
2	150	0.75	0.45
3	200	1.0	0.6

1.2.2. Inlet and Flow Boundary Conditions

Steady state flow conditions were assumed for the flow through the pipes. The range of flushing velocities studied was from 0.4m/s to 1.4m/s, to encompass flush rates currently used, for all of the pipe configurations.
The inlet conditions were velocity profiles normal to the inlet faces. When a flat inlet profile is used, it requires many pipe diameters for an accurate velocity profile to develop. Extra models were therefore used to simulate the flow profile after 48 pipe diameters. These profiles were then used to create accurate inlet profiles for the smooth, slightly tuberculated and tuberculated models.

The flow was modelled using the k-ε turbulence model and standard wall functions to describe the boundary layer. The flow solution used second order discretisation of the momentum and turbulence terms to improve accuracy with the unstructured grid. The outlet boundary condition was a mass balance condition.

1.3. THEORY

A particle at rest on a surface across which a fluid is flowing will experience a shear force due to the velocity gradient close to the surface. This force will tend to lift the particle perpendicularly away from the surface. This lift force due to shear, F_s, may be expressed as (Saffman 1965):

$$F_s = K_s \frac{\mu V a^2 \tau^{1/2}}{\nu^{1/2}}$$

Where μ = viscosity (kg.m^{-1}.s^{-1})
V = relative velocity (m.s^{-1})
a = radius of spherical particle (m)
τ = velocity gradient in radial direction (s^{-1})
$\nu = \mu/\rho$ = kinematic viscosity (m^2.s^{-1})
K_s = Saffman constant

In order to lift the particle the shear force, F_s, must exceed the weight, W_p, of the particle. Hence, for a critical condition:

$$F_s \geq W_p = \frac{4\pi a^3}{3} \rho . g$$

Where $\rho = \rho_p - \rho_w$ = particle density relative to fluid (kg.m^{-3})
g = acceleration due to gravity (m.s^{-2})

Substituting for F_s and rearranging gives:

$$V\tau^{1/2} \geq \frac{1}{K}\left(\frac{4\pi}{3}\right)\left(\frac{(\rho_p - \rho_w)}{\mu}.v^{1/2}\right).a$$

where K=K$_s$/g, and a is the particle radius measured in μm. For water this simplifies to:

$$V\tau^{1/2} \geq \frac{1}{K}.\frac{4\pi}{3}.(\rho_p - 1{,}000)a \qquad \text{Eqn. 1}$$

Where V is now the fluid velocity passing the particle at rest. The $V\tau^{1/2}$ term describes the flow conditions affecting the lift force due to shear in the near-wall layer.

Since the lifting of loose, stationary particles was considered, the relative velocity between the fluid and a particle equals the local fluid velocity. The local fluid velocity, V, and the local velocity gradient in the radial direction, τ, were calculated using the CFD model. The model was used to calculate these local values at the centres of the cells adjacent to the pipe wall. The distance from the cell centres to the wall varied for the different pipes, and approximate mean values are shown in Table B.3, below.

Table B.3
Wall and flow properties for the smooth pipes

Pipe	Approx. mean distance of cell centres from wall (μm)	Reynolds number, Re	y*
4"	850-1000	40,000 – 140,000	20 - 70
6"	1275-1500	60,000 – 210,000	26 - 100
8"	1700 – 2000	80,000 – 280,000	30 - 120

Due to the computational limits on the number of cells in the model, the wall-cells were larger than the particles that were studied, and hence their cell centres were further from the pipe wall than the particles. Therefore, to establish the local values (V & τ) for the particles, it was necessary to interpolate the velocity and velocity gradient between the wall cell centre and the wall, see **Figure B.2**.

The velocity was interpolated using the log law function. The validity of the log law was tested using the dimensionless property y*, which represents a length scale within the boundary layer [(2)]. The log law is known to be valid for values of y* between 30 and 60. This was assumed to be valid for all of the flows used.

To interpolate the velocity gradient, values of *dV/da* (= τ) from the CFD model were plotted as a linear function of *1/a*.

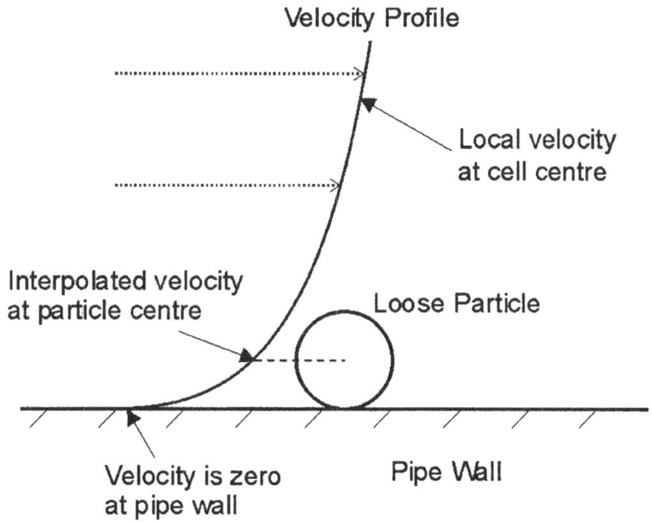

Figure B.2. Interpolation of local velocity values

The local velocity and velocity gradient profiles were calculated for each of the three pipe diameters. For each pipe diameter, these profiles were assumed to be valid for all states of tuberculation, a necessary assumption due to mesh size limitations.

The lift term, $V\tau^{1/2}$, was then calculated from the interpolated local values for a range of pipe diameters, flow rates, particle sizes and densities. The critical condition equation, Eqn. 1, was used to determine whether a particle would be removed.

The weight always acts vertically whereas the lift term acts radially towards the pipe axis. The critical lift condition is a worst case, and applies at the bottom centre line of the pipe, where the lift and weight terms act in opposite directions. For a smooth pipe the lift terms may be corrected to account for the increasing angle of the pipe wall, away from the centre line. However for the tuberculated pipes, the angle of the pipe wall is not known at any point and therefore no angle correction was used for any of the pipes. This assumes that all locations on the pipe wall behave as the worst case.

The value of the constant, K, was determined experimentally by trials on a pipe test rig, the results of which are summarised in **Table B.4**.

1.4. PARTICLES

The particles used in the model were assumed to be spherical. The particle sizes and densities used for each pipe diameter, tuberculation code and flow rate were in keeping with current available literature, and are shown in Table B.3. Data collected, to date, from mains in Thames Water have an average specific gravity of 3.03.

Table B.3
Particle properties

Specific gravity	Density (kg.m^{-3})
1.1	1,100
1.5	1,500
2.0	2,000
3.0	3,000

Particle diameter (μm)
20
50
120
200

When evaluating the lift term, the local velocity and velocity gradient are calculated for the particle centre. So for a particle of diameter 50 μm, the local values are calculated at a distance of 25 μm from the pipe wall.

1.5. DERIVATION OF K

The lifting criterion has been developed by equating the lift force with the weight of the particle (Eqn 1). The constant, *K*, however was unknown for the flow conditions being studied. Pipe-rig trials were therefore carried out to establish a value of K. These trials were reported in the third periodic report.

The trials used a 4" smooth pipe that was seeded using particles of known sizes and densities. The trial results established the mean pipe velocity required to just lift the particles. For this known pipe velocity, the $V\tau^{1/2}$ term was calculated from the CFD simulation for the near-wall region. Plotting:

$$V\tau^{\frac{1}{2}} \quad against \quad \frac{4\pi}{3}.(\rho_p - 1{,}000)a$$

gives a straight line, the gradient of which is *1/K*, where *a* is measured in μm.

1.6. LIFT TERM CALCULATION

For each case, the lift term was calculated for every wall-cell in the tuberculated region of the pipe invert. These values were exported from the CFD package and collected in a database. The values of the right hand side of the critical lift condition (Eqn. 1) were also calculated for each particle size and density.

The database was then used to calculate the number of cells with lift terms higher than the relevant critical value, and hence the fractional pipe length cleaned.

2. RESULTS

2.1. FLOW

Typical flow results are shown for the smooth, slightly tuberculated and tuberculated pipes in Figures B.5, B.6 and B.7 respectively. The views shown are typical sections of the bottom halves of the pipes, viewed from above, with an example mean velocity of 1.0 m/s. The pipe walls are coloured by the local velocity magnitude at the wall cell centres. The arrows show local velocity vectors. Each arrow shows the flow direction and its length is proportional to the velocity magnitude.

The near-wall flow in the smooth pipe is parallel to the pipe axis and is even around the pipe wall. The legend shows the range of velocities across the whole pipe. The velocity values vary from 1.17 m/s (along the pipe axis) to 0.72 m/s (near the wall).

In the slightly tuberculated pipe, the flow is mainly even and axial with small perturbations around the tubercles. The flow immediately downstream of the tubercles is less than the mean wall flow. This increases the range of velocities, with the lowest velocity at 0.045 m/s.

In the tuberculated pipe, the flow is highly influenced by the tubercle structure with higher flows across the tops and lower flows in the spaces between. There are significant near wall flows that are not parallel to the pipe axis. The range of velocities is much increased with the highest velocity at 2.91 m/s, which is significantly higher than the mean pipe flow. The lowest flows are around 0.000 0016 m/s, which is effectively zero and these stagnant regions occur in the small spaces between tubercles.

2.2. NEAR-WALL VELOCITY PROFILES

Local velocity magnitude and gradient values close to the wall were interpolated using the wall-cell centre values. For the smooth pipes, this procedure was carried out for each mean velocity, and for each pipe diameter. These functions were found to give a good fit for the data.

For the example case shown in **Figure B.3**, the data are fitted with the following equations:

$$V = 0.062 \times \ln(a) + 0.355$$

$$\tau = \frac{dV}{da} = \frac{116{,}656}{a}$$

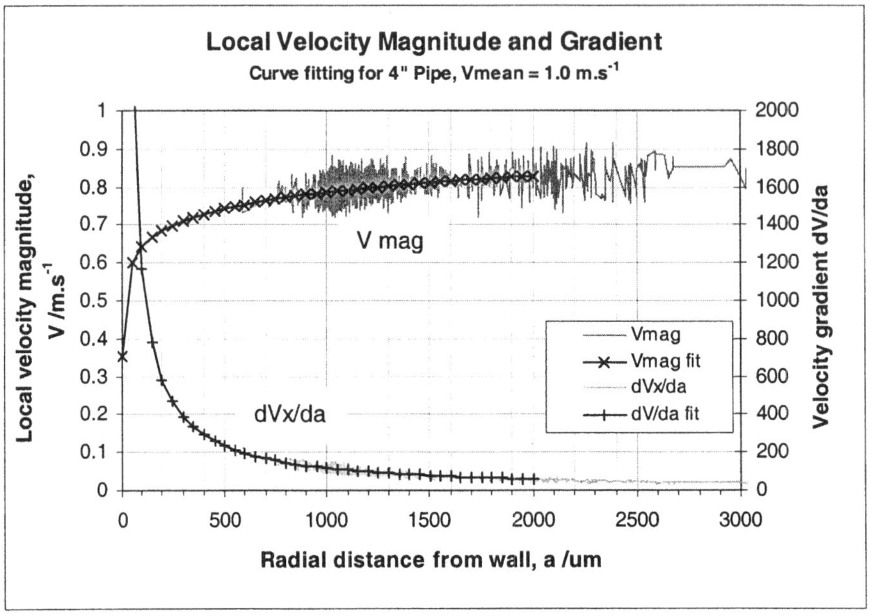

Figure B.3. Curve fitting for velocity magnitude and gradient in the near-wall region

2.3. DERIVATION OF K

Figure B.4 shows the results of the curve fitting to calculate *K*. The axis labels LHS and RHS refer to the left and right hand sides of Eqn. 1, without *K*. The data points are from the pipe rig trials for sand, GAC and distribution particles.

The error bars are based upon a ±10% error in the flow measurements and a ±33% error in the particle sizes due to the use of sieve fractions.

Table B.4, below summarizes results from the pipe test rig trials on lifting velocities of particles of different size and density.

Table B.4
Summary of pipe rig trial results

Material	Pipe diameter inches (mm)	Particle specific gravity	Particle diameter (μm)	Flow to remove ($m.s^{-1}$)
Sand	4" (100)	2.73	0 – 275	0.55 – 0.58
GAC	4" (100)	1.65	0 - 1,400	0.45 – 0.52
SiO_2	4" (100)	2.43	250	0.65
Distributio n	4" (100)	2.83	0 - 1,180	0.48 – 0.72

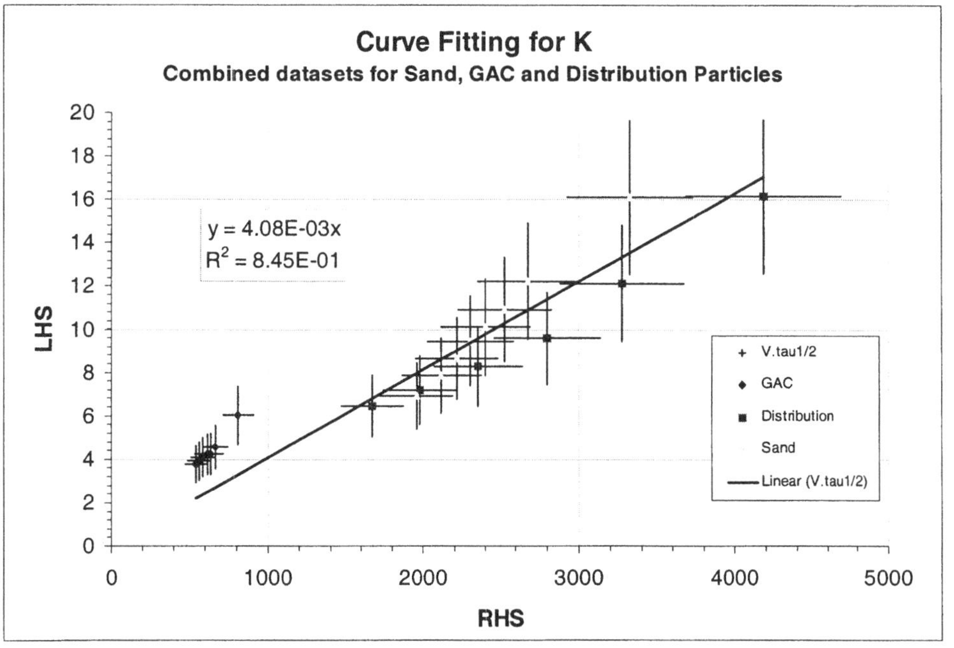

Figure B.4. Fitting pipe-rig trial data to calculate K

The best straight-line fit of the data was found using $a^{-0.25}$. This gives a value of $K = 1/0.00408 = 245$. Substituting into Eqn.1 gives the following critical lift condition:

$$V\tau^{1/2} \geq 0.017(\rho_p - 1{,}000)a^{-0.25} \qquad \text{Eqn. 2}$$

2.4. CALCULATION OF LIFT CRITERIA

Eqn. 2 can now be used to calculate specific particle removal criteria. For example, particles of size 120μm and specific gravity 2.5 will be lifted when:

$$V\tau^{1/2} \geq 0.017(2{,}500 - 1{,}000)60^{-0.25}$$

$$V\tau^{1/2} \geq 9.16$$

Table B.6 shows the calculated lift criteria for all of the particles studied.

3. REFERENCES

1. Saffman P. G., *The Lift on a Small Sphere in a Slow Shear Flow,* J. Fluid Mech., 22:385--400, 1965.

2. B. E. Launder and D. B. Spalding, *The Numerical Computation of Turbulent Flows*, Computer Methods in Applied Mechanics and Engineering, 3:269--289, 1974.

APPENDIX C
PILOT STUDY FLUSHING METHODOLOGY

APPENDIX C
PILOT STUDY FLUSHING METHODOLOGY

Appendix C1: Methodology for iron sludge covered pipe trials

The methodology below was followed for each of the iron test pipe samples used, which included:

- 1 x iron pipe section from the stagnation (intermittent flow) section of the TORUS pipe rig; and
- 2 x iron pipe sections from the continuous flow section of the TORUS pipe rig.

Insert and incubate pipe

1. *Photograph pipe* - the crown and invert at each end of the pipe were photographed.
2. *Insert pipe* - the pipe was inserted and bolted into the rig between two flanges (note: upstream and downstream of both flanges was clear PVC).
3. *Start pump* - the pump was started with the hydrant slightly open to allow only a low flow through the rig. The valves were opened to release air from the rig. The rig was filled.
4. *Incubate pipe* - the pump was turned off and the pipe was left in the filled rig to wet and incubate for up to 6 days.

Perform trial

5. *Tie hydrant net* - the hydrant net was tied onto the end of a hose attached to the hydrant.
6. *Start pump* - the pump was started with the hydrant slightly open to allow only a low flow through the rig. The valves were opened to release air from the rig.
7. *Step flow increase* - the flow rate was set to the first trial flow of 1 L/s.
8. *Flush for 5 minutes* - the pipe rig was flushed at the trial velocity of 1 L/s for a test duration of approximately 5 minutes.
9. *Stop flow* - the flow through the rig was stopped temporarily, to enable removal of the hydrant net from the end of the hose.
10. *Examine hydrant net* - the hydrant net was removed, and its interior was examined for deposits (i.e.: material dislodged and transported during the experiment). Any particles captured were collected in a labelled container.
11. *Clean hydrant net* - the hydrant net was cleaned and re-tied onto the end of the hose.
12. *Step flow increase* - the flow rate was increased to the second trial flow of 2 L/s.
13. *Flush for 5 minutes* - the pipe rig was flushed at the trial velocity of 2 L/s for a test duration of approximately 5 minutes.
14. *Stop flow* - the flow through the rig was stopped temporarily, to enable removal of the hydrant net from the end of the hose.
15. *Examine hydrant net* - the hydrant net was removed, and its interior was examined for deposits (ie: material dislodged and transported during the experiment). Any particles captured were collected in a labelled container.
16. *Clean hydrant net* - the hydrant net was cleaned and tied onto the end of the hose.

17. *Repeat steps up to maximum flow (i.e.: 22 L/s)* - the above steps were then repeated (i.e.: increasing the flow by 1 L/s for each trial, flushing the pipe rig for approximately 5 minutes, removing and examining and replacing the hydrant net, and collecting captured particles as applicable).
18. *Stop pump* - the pump was then turned off, and the hydrant was closed.
19. *Remove test pipe* - the section of test pipe was then removed from the rig. The crown and invert at each end of the pipe was then re-photographed for comparison with the pre-flush photographs.
20. *Particle analysis* - the particles were then analysed, which involved recording whether particles were present in the hydrant net at each flow rate.

Appendix C2: Methodology for manganese coated pipe flushing trials

Pipes were obtained from the distribution system in the Netherlands.

Install and incubate pipe

1. *Photograph pipe* - the crown and invert at each end of the pipe was photographed.
2. *Insert pipe* - the pipe was inserted and bolted into the rig between two flanges (note: upstream and downstream of both flanges was clear PVC).
3. *Start pump* - the pump was started with the hydrant slightly open to allow only a low flow through the rig. The valves were opened to release air from the rig. The rig was filled.
4. *Incubate pipe* - the pump was turned off and the pipe was left in the filled rig to wet and incubate.

Perform trial

5. *Tie hydrant net* - the hydrant net was tied onto the end of the hose.
6. *Start pump* - the pump was started with the hydrant slightly open to allow only a low flow through the rig. The valves were opened to release air from the rig.
7. *Step flow increase* - the flow rate was set to the first trial flow of 1 L/s.
8. *Flush for 5 minutes* - the pipe rig was flushed at the trial velocity of 1 L/s for a test duration of approximately 5 minutes.
9. *Stop flow* - the flow through the rig was stopped temporarily, to enable removal of the hydrant net from the end of the hose.
10. *Examine hydrant net* - the hydrant net was removed, and its interior was examined for deposits (i.e.: material dislodged and transported during the experiment). Any particles captured were collected in a labelled container.
11. *Clean hydrant net* - the hydrant net was cleaned and re-tied onto the end of the hose.
12. *Step flow increase* - the flow rate was increased to the second trial flow of 2 L/s.
13. *Flush for 5 minutes* - the pipe rig was flushed at the trial velocity of 2 L/s for a test duration of approximately 5 minutes.
14. *Stop flow* - the flow through the rig was stopped temporarily, to enable removal of the hydrant net from the end of the hose.

15. *Examine hydrant net* - the hydrant net was removed, and its interior was examined for deposits (i.e.: material dislodged and transported during the experiment). Any particles captured were collected in a labelled container.
16. *Clean hydrant net* - the hydrant net was cleaned and tied onto the end of the hose.
17. *Repeat steps up to maximum flow (ie: 22 L/s)* - the above steps were then repeated (i.e.: increasing the flow by 1 L/s for each trial, flushing the pipe rig for approximately 5 minutes, removing and examining and replacing the hydrant net, and collecting captured particles as applicable).
18. *Stop pump* - the pump was then turned off, and the hydrant was closed.
19. *Remove test pipe* - the section of test pipe was then removed from the rig. The crown and invert at each end of the pipe was then re-photographed for comparison with the pre-flush photographs.
20. *Particle analysis* - a laboratory elemental analysis was then performed on the particles captured.

Appendix C3: Methodology for GAC flushing trials

The methodology below was followed for each of the GAC trials performed, which included:

- 3 trials with GAC seeded into an iron pipe; and
- 2 trials with GAC seeded into a PVC pipe.

The GAC was obtained from filter beds at a water treatment plant operated by Thames Water in southwest London.

Insert GAC and incubate pipe

1. *Insert test pipe* - the section of test pipe (iron or PVC) was inserted and bolted into the rig between two flanges (note: upstream and downstream of both flanges was clear PVC).
2. *Start pump* - the pump was started with the hydrant slightly open to allow only a low flow through the rig. The valves were opened to release air from the rig. The rig was filled.
3. *Insert GAC* - the pump was left running at a minimal rate, and the GAC was inserted into the rig using a 2-valved tapping point. The tapping was point was upstream of the iron pipe and a small flow was therefore required to transport the GAC downstream so it settled in the test pipe. When the PVC pipe was used, the tapping point was in the centre of the test pipe and zero flow conditions were used when inserting the GAC.
4. *Incubate GAC* - the pump was turned off and the GAC was left in the filled rig to wet and incubate for up to 5 days.

Perform trial

5. *Tie hydrant net* - the hydrant net was tied onto the end of the hose.
6. *Start pump* - the pump was started with the hydrant slightly open to allow only a low flow through the rig. The valves were opened to release air from the rig.
7. *Gradually increase flow* - the flow rate was gradually increased over a period of several minutes. The flow was increased until the maximum capacity of the pump was reached or all

material was observed to have been removed. Any transportation of particles through the rig was photographed. The flows at which any particles were transported through the rig were recorded.

8. *Stop flow* - the flow through the rig was stopped
9. *Examine hydrant net* - the hydrant net was removed, and its interior was examined for deposits (i.e.: material dislodged and transported during the experiment). Any particles captured were collected in a labelled container.
10. *Clean hydrant net* - the hydrant net was cleaned and re-tied onto the end of the hose.
11. *Particle analysis* - the particles were then analysed, which involved commenting on the amount of GAC captured in the net (i.e.: compared to the amount inserted into the rig through the tapping point).

Appendix C4: Methodology for ferric floc flushing trials

The methodology below was followed for each of the ferric floc trials performed, which included:

- 2 trials with ferric floc deposited in an iron pipe; and
- 1 trial with ferric floc deposited in a PVC pipe.

Make ferric floc

1. *Collect raw water* - raw water samples were collected from the on-site canal. Raw water was used instead of filtered water to increase the amount of floc produced.
2. *Make ferric stock solution* - a stock solution was made with 40% strength ferric sulphate (14mL per 1 litre of deionised water).
3. *Perform jar tests to make floc* - the raw water was dosed with the stock in jar test equipment and jar tests were performed until sufficient floc was formed (floc from approximately 12 litres of raw water was used per trial).
4. *Collect and settle floc* - after each jar test was performed the clear water was decanted off and the floc was collected in a sample jar. The floc was then allowed to settle for several days. After several days, any clear water above the floc was decanted off.

Insert floc and incubate pipe

5. *Insert test pipe* - the section of test pipe (iron or PVC) was inserted and bolted into the rig between two flanges (note: upstream and downstream of both flanges was clear PVC).
6. *Start pump* - the pump was started with the hydrant slightly open to allow only a low flow through the rig. The valves were opened to release air from the rig. The rig was filled.
7. *Stop pump* - the flow through the rig was stopped after the rig was filled. There could not be any flow through the rig when the floc was inserted because the floc was too light and would not settle on the base of the pipe unless there were zero flow conditions.
8. *Insert ferric floc* - the floc was inserted into the rig using a 2-valved tapping point. The tapping was point was upstream of the iron pipe, and was in the centre of the PVC pipe. Zero flow conditions were used when feeding the floc into both pipes.
9. *Incubate floc* - the floc was left in the filled rig to incubate for up to 4 days

Perform trial

10. *Tie hydrant net* - the hydrant net was tied onto the end of the hose.
11. *Start pump* - the pump was started with the hydrant slightly open to allow only a low flow through the rig. The valves were opened to release air from the rig.
12. *Gradually increase flow* - the flow rate was gradually increased over a period of several minutes. The flow was increased until the maximum capacity of the pump was reached or all the material was observed to have been removed. Any transportation of particles through the rig was photographed. The flows at which any particles were transported through the rig were recorded.
13. *Stop flow* - the flow through the rig was stopped.
14. *Examine hydrant net* - the hydrant net was removed, and its interior was examined for deposits (i.e.: material dislodged and transported during the experiment). Any particles captured were collected in a labelled container.
15. *Clean hydrant net* - the hydrant net was cleaned and re-tied onto the end of the hose.
16. *Particle analysis* - any particles captured were then analyzed, which involved commenting on the presence or absence of ferric floc in the hydrant net.

Appendix C5: Methodology for Adherent Deposit Flushing Trials

The method followed to perform adherent deposit flushing trials is described below:

1. *Photograph pipe* - the crown and invert at each end of the pipe was photographed.
2. *Insert pipe* - the pipe was inserted and bolted into the rig between two flanges.
3. *Start pump* - the pump was started with the hydrant closed and the valves open to release air from the rig.
4. *Open hydrant slightly* - the hydrant was opened slightly to provide a slow flow through the system. The system was flushed for 1 minute to remove any loose deposits in the rig or on the pipe wall.
5. *Tie hydrant net* - the hydrant net was tied onto the end of the re-circulating hose (i.e.: the hose that feeds water back into the tank to re-circulate the water through the system).
6. *Step flow increase* - the flow rate was increased to the first trial flow of 1 L/s.
7. *Flush for 5 minutes* - the pipe rig was flushed at the trial velocity of 1 L/s for a test duration of 5 minutes.
8. *Stop flow* - the flow through the rig was stopped temporarily, to enable removal of the hydrant net from the end of the re-circulating hose.
9. *Examine hydrant net* - after the hydrant net was removed, it's interior was examined for deposits (i.e.: material dislodged and transported during the experiment). Any corrosion deposits were collected in a labelled container.
10. *Clean hydrant net* - a clean hydrant net was tied onto the end of the re-circulating hose.
11. *Step flow increase* - the flow rate was increased to the second trial flow of 2 L/s.
12. *Flush for 5 minutes* - the pipe rig was flushed at the trial velocity of 2 L/s for a test duration of 5 minutes.
13. *Stop flow* - the flow through the rig was stopped temporarily, to enable removal of the hydrant net from the end of the re-circulating hose.

14. *Examine hydrant net* - after the hydrant net was removed, it's interior was examined for deposits (i.e.: material dislodged and transported during the experiment). Any corrosion deposits were collected in a labelled container
15. *Clean hydrant net* - a clean hydrant net was tied onto the end of the re-circulating hose.
16. *Repeat steps up to maximum flow* - the above steps were then repeated, ie: increasing the flow by 1 L/s for each trial, flushing the pipe rig for 5 minutes, removing and examining and replacing the hydrant net, and collecting deposits as applicable.
17. *Stop pump* - the pump was then turned off, and the hydrant was closed.
18. *Remove test pipe* - the section of test pipe was then removed from the rig. The crown and invert at each end of the pipe was then re-photographed for comparison with the pre-flush photographs.
19. *Particle analysis* - the particles were then analysed, which involved recording whether particles were present in the hydrant net at each flow rate, photographing any particles collected (using a microscope with a camera attached), and dry weighing the particles if sufficient quantities were collected.
20. *Repeat process with new pipe section* - the above method was followed for each of the test pipe samples used.

APPENDIX D
EXAMPLE FIELD PROTOCOL FOR UNIDIRECTIONAL FLUSHING

APPENDIX D
EXAMPLE FIELD PROTOCOL FOR UNIDIRECTIONAL FLUSHING

AWWARF – Project # 2606
Dover, New Hampshire
Date: July 2000

Revision No.:

Revision Date: ____________________

The target velocities of approximately 2 – 4, and at least 6 ft/s can be achieved for this loop by flushing the designated hydrants through either of the 2 ½-inch ports. Residual pressures should be monitored throughout flushing to ensure that the pressure will not drop below 20 psi.

Step 1 - *Close V1094*
Close V1099
Close V1085
Close V1009
Close V1079
Close V1254
Close V1233
Prepare hydrant FH588 for sampling task 3 (velocity & time-to-clear)

Open FH588
Wait until discharged water clears at FH588. Measure and record velocity and time-to-clear (sampling task 3).

Step 2 - *Close FH588*
Prepare hydrant FH560 for sampling task 3 (velocity & time-to-clear)

Open FH560
Wait until discharged water clears at FH560. Measure and record velocity and time-to-clear (sampling task 3).

Step 3 - *Close FH560*
Close V1007
Close V1014
Prepare hydrant FH556 for sampling task 2 (water quality parameters)

Open FH556

Follow instructions for sampling task 2 (water quality parameters). This hydrant will be flushed twice at two different velocities during this step.

Step 4 - *Close FH556*
Close V1170
Prepare hydrant FH598 for sampling task 3 (velocity & time-to-clear)

Open FH598
Wait until discharged water clears at FH598. Measure and record velocity and time-to-clear (sampling task 3).

Step 5 - *Close FH598*
Open V1094
Close V1171
Prepare hydrant FH593 for sampling task 3 (velocity & time-to-clear)

Open FH593
Wait until discharged water clears at FH593. Measure and record velocity and time-to-clear (sampling task 3).

Step 6 - *Close FH593*
Prepare hydrant FH596 for sampling task 3 (velocity & time-to-clear)

Open FH596
Wait until discharged water clears at FH596. Measure and record velocity and time-to-clear (sampling task 3).

Step 7 - *Close FH596*
Open V1254
Close V1078
Prepare hydrant FH598 for sampling task 3 (velocity & time-to-clear)

Open FH598
Wait until discharged water clears at FH598. Measure and record velocity and time-to-clear (sampling task 3).

Step 8 - *Close FH598*
Open FH596
Wait until discharged water clears at FH596 (no sampling).

Step 9 - *Close FH596*
Open V1079
Open V1085
Close V1094
Close V1087

Prepare hydrant FH595 for sampling task 2 (water quality parameters)

Open FH595
Follow instructions for sampling task 2 (water quality parameters). This hydrant will be flushed twice at two different velocities during this step.

Step 10 *- Close FH595*
Open V1014
Close V1017
Open FH556
Wait until discharged water clears at FH556 (no sampling).

Step 11 *- Close FH556*
Open V1017
Prepare hydrant FH594 for sampling task 1 (particle characterization). See sampling protocol for specific instructions. This hydrant will be flushed at two different velocities during this step.

Open FH594

Step 12 - Close FH594
Open V1170
Close V1254
Prepare hydrant FH598 for sampling task 3 (velocity & time-to-clear)

Wait until discharged water clears at FH598. Measure and record velocity and time-to-clear (sampling task 3).
Open FH598

Step 13 - Close FH598
Open V1171
Open V1254
Open V1078
Open V1094
Prepare hydrant FH597 for sampling task 3 (velocity & time-to-clear)

Wait until discharged water clears at FH597. Measure and record velocity and time-to-clear (sampling task 3).
Open FH597

Step 14 - Close FH597
Open V1087
Close V1085
Open FH591

Wait until discharged water clears at FH591 (no sampling).

Step 15 - Close FH591
Open V1007
Close V1087
Prepare hydrant FH590 for sampling task 3 (velocity & time-to-clear)

Wait until discharged water clears at FH590. Measure and record velocity and time-to-clear (sampling task 3).
Open FH590

Step 16 - Close FH590
Open V1085
Open V1099
Open V1009
Open V1233
Open V1087

Sampling Protocol for AWWARF#2606 "Establishing Contaminant Flushing Velocities"

1 OVERVIEW

Not all the utilities that participate in the study will have the same sampling and analytical requirements. Therefore, separate tasks were established as shown in Table 1. The specific hydrants where sampling tasks 1 and 2 have to be conducted are indicated in the flushing protocol. Optionally, the utility may sample additional hydrants. Detailed procedures for sampling are contained in the subsequent sections. Utility specific water quality parameters are listed in Table 2. They were selected based on the water quality problem(s) described by the utility. Sample analysis should be conducted in accordance with the listed Standard Method[1]. If other analytical methods are used, the utility should indicate the method in the sample report/flushing record form.

Completed sample reports/flushing forms including the results from laboratory analyses should be mailed to:

Utility	Manager
Cincinnati, Cleveland, Irvine Ranch, Newport News, Portland, Southern California Water Company	Peter Pommerenk AH Environmental Consultants, Inc 11828 Canon Boulevard, Suite H Newport News, VA 23606 Phone (757) 873-4959
Bellevue, Dover, South Berwick, Vancouver, Yarra Valley Water	Kathy Martel Economic and Engineering Services, Inc. 581 Pequawket Trail Steep Falls, ME 0408 Phone (207) 675-3437

[1] Standard Methods for the Examination of Water and Wastewater. American Public Health Association, Washington, D.C., 19th edition, 1995

Table 1. Sampling tasks.

SAMPLING TASK	UTILITY	WHERE	FLUSHING VELOCITY	# OF SAMPLES	PARAMETER	ANALYSIS BY
1	All participating utilities except Southern CA, Vancouver and Yarra Valley	At least one selected hydrant	At least two velocities	One sample per velocity and hydrant	Particle characterization (hydrant nets, shipped out for analysis) velocity, time-to-clear (on-site)	MSU, Thames Utility
2	All participating utilities	At least one hydrant (other than those sampled in sampling task 1)	Two velocities	(At least) three timed samples per velocity	Bulk water pH, Temp., Turbidity, disinfectant residual (on-site) + utility specific parameters (lab) + velocity, time-to-clear	Utility
3	All participating utilities	All remaining hydrants not sampled in sampling task 1 or 2	Maximum velocity	N/A	Velocity, time-to-clear	Utility

Table 2. Utility specific sampling parameters.

	Utility	Bellevue	Cincinnati	Cleveland	Dover	Irvine Ranch	Newport News	Portland	South Berwick	Southern CA	Vancouver	Yarra Valley	Standard Method #
ON-SITE	PH	X	X	X	X	X	X	X	X	X		X	4500-H^+ B.
	Temperature	X	X	X	X	X	X	X	X	X		X	2550 B.
	Disinfectant residual	X	X	X	X	X	X	X	X	X		X	4500-Cl G.
	Turbidity	X	X	X	X	X	X	X	X	X		X	2130 B.
LABORATORY ANALYSIS	Total reactive phosphorus			X									4500-P E.
	Total Iron[2]	X	X	X	X	X	X	X	X	X		X	3500-Fe B or C
	Total Manganese[3]				X				X				3500-Mn A. or B.
	Total Hardness	X	X	X	X				X				2340 C.
	Alkalinity	X	X	X	X				X				2320 B.
	HPC	X	X	X	X		X	X	X	X		X	9215 (R2A-agar)
PARTICLE SAMPLES	Chemical Analysis	Fe, Ca	Fe, Ca	Fe, Ca, PO_4	Fe, Mn, Ca	Fe	Fe	Fe	Fe, Mn, Ca	N/A		N/A	Samples shipped to and analyzed by MSU
	HPC + dry weight analysis (on samples from one hydant only)	X	X	X	X		X	X	X	N/A		N/A	

[2] Sample digestion is required (Standard Method # 3030E).
[3] Acidification and/or digestion required.

2 TASK 1: PARTICLE CHARACTERIZATION

2.1 OBJECTIVE

To collect different types of particles and deposits for characterization under known controlled flow conditions.

2.2 SITE INFORMATION

For each hydrant the following data should be obtained;

- when the hydrant was last flushed,
- the maximum and average daily flow rate in the main to be flushed,
- the length of pipe to be flushed must be calculated and a checked against main maps carried out to ensure that the diameter is the same along the length and to determine pipe volume.
- other information as specified on the flushing record form

Notes:

- The flushing protocol involves flushing at low and high velocities. Some parts of a distribution system may not be able to support a high rate flush. To avoid pressure or supply problems it is essential that the protocol is signed off by the local network operators department and communication with local operations maintained throughout flushing exercise.
- Hydrant flushing has the potential to generate extreme discoloration in nearby properties. Although transient, it may cause adverse customer reaction. For this reason, local customers and health authorities should be notified at least one week prior to hydrant flushing.

2.3 METHODOLOGY :

1. Attach a pressure gauge to the hydrant and take a reading (static pressure).
2. Gently crack open the hydrant and flush for 1 minute to clear the hydrant bowl and the feeder main.
3. Attach a flow meter (or pitot gauge) to the hydrant and secure a twin hydrant net, with the coarse net (308 microns) on the inside and the fine net (57 microns) on the outside, onto the end of the flow meter or pitot gauge.

4. Set the flow rate to achieve a velocity of 2 – 4 feet per second. The appropriate flow rates or pitot pressures can be determined using the flushing form.
5. Flush the hydrant at this flow rate until the discharged water has cleared AND a minimum of one complete pipe volume has been discharged. **IF MORE THAN APPROXIMATELY ONE POUND OF MATERIAL HAS ACCUMULATED IN THE TWIN NET, CLOSE THE HYDRANT, REMOVE THE NET, AND FINISH FLUSHING. RECORD ACTUAL TIME OF FLUSH.**
6. After the flush, close the hydrant and remove the hydrant nets and store nets in sealed plastic bag labelled with date, location and flow rate.
7. Secure another twin hydrant net (coarse net on inside, fine net on outside) onto the end of the flow meter.
8. Open the hydrant, set flow rate to obtain a velocity of greater than 6 feet per second. The required pitot pressures can be determined using the flushing form. If it was not possible to achieve the target flow rate, then please note the flow rate achieved on the samples.
9. Flush the hydrant at this flow rate until the discharged water clears AND a minimum of one complete pipe volume has been discharged
10. After the flush, close the hydrant and remove the hydrant nets and store nets in sealed plastic bag labelled with date, location and flow rate.

Notes:

- The flushing of very dirty mains may result in the nets bursting. This happens either because the net becomes full or clogged or because fine net is torn from abrasive action as a result of contact with the rough road surface. The clogging of the nets must be avoided since fine particles will become trapped. Practitioners stroke the net during flushing (known as 'milking the net'!!) to avoid particles becoming clogged. The fine net must be protected from the abrasive road surface. The hydrant net can be surrounded by a tube to ensure it is not resting on a rough surface.

11. The hydrant nets in the sealed bags have to be shipped in coolers overnight to:

Anne Camper
Department of Civil Engineering
205 Cobleigh Hall
Montana State University-Bozeman
Bozeman, MT 59717
Phone: (406) 994-4906

3 TASK 2: BULK WATER QUALITY PARAMETERS

3.1 Objective

To determine changes in water quality due to flushing.

3.2 Site Information

For each hydrant the information as specified on the flushing record form should be obtained.

Notes:

- The flushing protocol involves flushing at low and high velocities. Some parts of a distribution system may not be able to support a high rate flush. To avoid pressure or supply problems it is essential that the protocol is signed off by the local network operators department and communication with local operations maintained throughout flushing exercise.
- Hydrant flushing has the potential to generate extreme discoloration in nearby properties. Although transient, it may cause adverse customer reaction. For this reason, local customers and health authorities should be notified at least one week prior to hydrant flushing.

3.3 Preparations

Ensure that the field crew has a sufficient number of prelabeled sample containers and steril sample bags (for HPC, if applicable) and that the appropriate test kits are available to conduct on-site analysis for pH, disinfectant residual, turbidity, and temperature.

3.4 Methodology :

12. Attach a pressure gauge to the hydrant and take a reading (static pressure).
13. Gently crack open the hydrant and flush for 1 minute to clear the hydrant bowl and the feeder main.
14. Attach a flow meter (or pitot gauge) to the hydrant. In pipe sections where two hydrants are opened for flushing no velocity measurements will be made.
15. Set the flow rate (or the pitot pressure) to achieve a velocity of 2 – 4 feet per second. The appropriate flow rates or pitot pressures can be determined using the flushing form. Flush the hydrant at this flow rate until the discharged water has cleared but for at least ten (10) minutes. The discharged water can be considered "clear" if the turbidity has decreased to the value measured at the clean source AND remained unchanged for two minutes. Alternative stopping criteria can be
    - The volume of the flushed water has exceeded discharge limitations, or

- The residual pressure falls below 20 psi.

If flushing was stopped before the discharged water cleared, indicate the reason on the flushing form.

16. During the 10-minute flush take at least three (3) grab samples for on-site analysis of
    - Temperature (Standard Method 2550 B),
    - pH (Standard Method 4500-H^+ B),
    - Disinfectant residual (Standard Method 4500-Cl G), and
    - Turbidity (Standard Method 2130 B).

    The samples should be taken prior to flushing (e.g. after clearing the hydrant barrel), when the discharged water appears most turbid, and after the water has cleared. Turbidity measurements are particularily important, because they are likely to correlate to other water quality parameters such as total iron and manganese. Therefore, the field crew should ensure that turbidity is monitored frequently during flushing.

17. During the 10-minute flush take at least three samples for laboratory analysis of the parameters specified in Table 2. The samples should be taken prior to flushing (e.g. after clearing the hydrant barrel), when the discharged water appears most turbid, and after the water has cleared. Record the time of sampling and the sample ID on the flushing form. For information on the appropriate sample technique and sample containers consult the utility's laboratory analyst.

18. After the first 10-minute flush, set flow rate to obtain a velocity of greater than 6 feet per second. The required pitot pressures can be determined using the flushing form. If it was not possible to achieve the target flow rate, then please note the flow rate achieved on the samples.

19. Flush the hydrant at this flow rate until the discharged water clears but for at least ten (10) minutes. The discharged water can be considered "clear" if the turbidity has decreased to the value measured at the clean source AND remained unchanged for two minutes. Alternative stopping criteria can be
    - The volume of the flushed water has exceeded discharge limitations, or
    - The residual pressure falls below 20 psi.

    If flushing was stopped before the discharged water cleared note the reason on the flushing form.

20. Collect samples according to 16. and 17.

AWWARF #2606 Flushing Record Form

1. Hydrant #	
2. Hydrant orifice diameter [in]	
3. Hydrant coefficient	
4. Pipe size [in]	
5. Length of pipe section	
6. Static pressure (psi)	
7. Target velocity [ft/s]	2-4 ft/s >6 ft/s
8. Required flow rate [gpm]	
9. Required pitot pressure [psi]	

Date:
Work performed by:

Flushing start time
Flushing stop time

10. Measured pitot pressure [psi]	
11. Time to clear [min]	

Loop
Step

12. Remarks:

Laboratory Samples (Sampling Task 2)		**On-Site Sample Analysis (Sampling Task 2 + 3)**				
Time	Sample ID	Time	Turbidity [NTU]	Temperature [°C]	Disinf. Res. [mg/L Cl_2]	pH

Notes:

1. *Hydrant #* Record the hydrant number as it appears in the flushing procedures.
2. *Hydrant orifice diameter* Measure and record the diameter of the orifice that is to be opened. ($D_{Orifice}$-inch).
3. *Hydrant coefficient* Determine the shape of the hydrant opening according to the figure below and record the appropriate hydrant coefficient C_D.

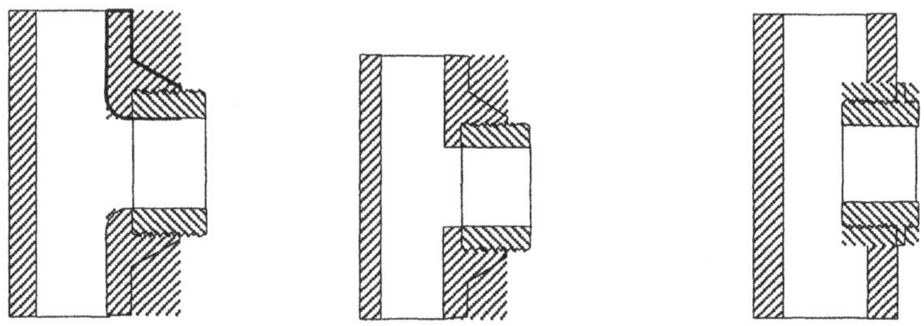

Smooth and Rounded
Coefficient = 0.90

Square and Sharp
Coefficient = 0.80

Square and Projecting
Coefficient = 0.70

4. *Pipe Size* Copy the size of the pipe (D_{Pipe}-inch) that is flushed from the flushing procedures.
5. *Length of pipe section* Record the length of the pipe section that is being flushed.
6. *Static pressure* Record the static pressure at the hydrant before flushing.
7. *Target Velocity* Circle the appropriate target velocity.
8. *Required flow rate* Determine the required flow rate Q from the table below

Pipe Diameter [in]	Target velocity [ft/s]	
	2 – 4	> 6
4	78 – 157	> 235
6	176 – 353	> 529
8	313 – 627	> 940

9. *Required outlet pressure* Using the required flow rate Q (gpm) , the orifice diameter $D_{orifice}$ (in), and the hydrant coefficient C_D, determine the required outlet (pitot) pressures P (psi) using the following formula:

$$P = (Q / (29.83 * C_D * D^2_{orifice}))^2$$

If the orifice diameter is 2 ½ inches, the table below can be used.

Required outlet (pitot) pressure for 2½-in orifice [psi]						
Pipe Size [in]	Target velocity 2 – 4 ft/s			Target velocity > 6 ft/s		
	$C_D = 0.7$	$C_D = 0.8$	$C_D = 0.9$	$C_D = 0.7$	$C_D = 0.8$	$C_D = 0.9$
4	0.4 – 1.4	0.3 – 1.1	0.2 – 0.9	> 3.2	> 2.5	> 2.0
6	1.8 – 7.3	1.4 – 5.6	1.1 – 4.4	> 16.4	> 12.6	> 9.9
8	5.8 – 23.1	4.4 – 17.7	3.5 – 13.9	> 51.9	> 39.7	> 31.4

10. *Measured pitot pressure:* Record the the actual pressure measured during flushing.

11. *Time-to-clear:* Record the time it takes until the turbidity of the discharged water does not change anymore or until the water appears clear.

12. *Remarks:* Note any observations made during flushing such as problems with the hydrant nets (if used), defective valves, hydrants, odor, visual observations (color), etc. Also, if known, record the date this hydrant was last flushed, the maximum daily flow rate in the main to be flushed.

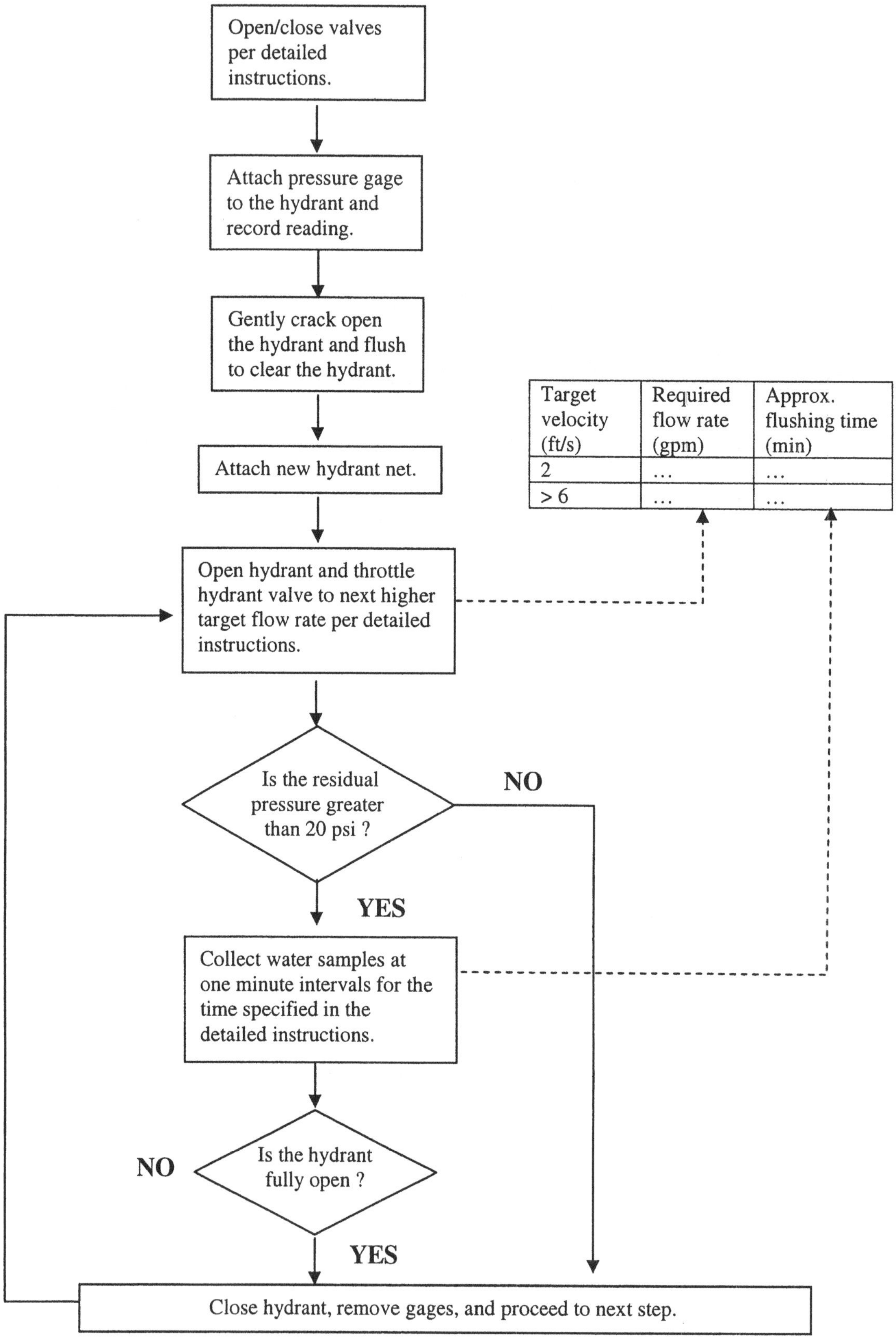
Open/close valves per detailed instructions.
Attach pressure gage to the hydrant and record reading.
Gently crack open the hydrant and flush to clear the hydrant.
Attach new hydrant net.
Open hydrant and throttle hydrant valve to next higher target flow rate per detailed instructions.
Target velocity (ft/s)
Required flow rate (gpm)
Approx. flushing time (min)
2
...
...
> 6
...
...
Is the residual pressure greater than 20 psi ?
NO
YES
Collect water samples at one minute intervals for the time specified in the detailed instructions.
Is the hydrant fully open ?
NO
YES
Close hydrant, remove gages, and proceed to next step.

APPENDIX E
STATISTICAL ANALYSIS OF FIELD STUDY RESULTS

APPENDIX E

STATISTICAL ANALYSIS OF FIELD STUDY RESULTS

Box-and-Whisker Plots

Box-and-whisker plotting is a powerful method of visualizing data and identifying central tendencies. The following discussion references the sample plot provided in Figure B.1. In these plots, the box encompasses the middle 50th percentile of data points. The endpoint values are referred to as the 25th and 75th percentile of the data, or the lower and upper quartile. The horizontal line within the box indicates the location of the median or 50th percentile. Circles at either end indicate the location of statistical outliers, which are points located more than 1.5 times away from the inter-quartile range (the difference between the 75th and 25th percentile). The vertical lines above and below the box, referred to as whiskers, extend to the largest and smallest values that are not classified as statistical outliers.

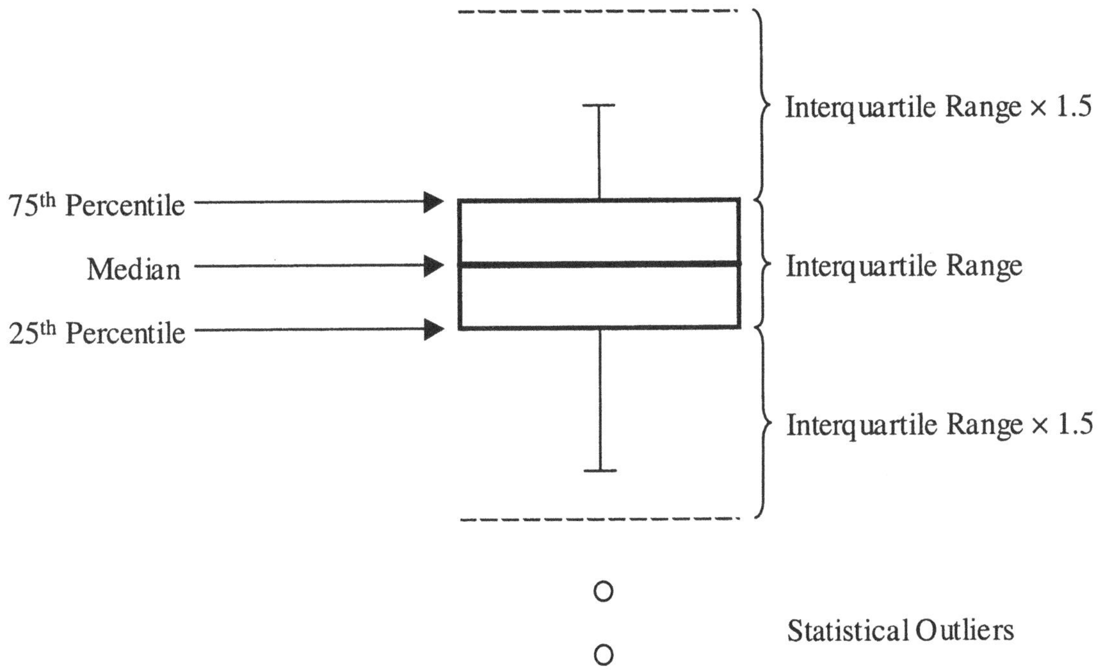

Figure E.1 Box-and-whisker plot example

Box-and-whisker plots (box-plots) were generated to determine the impact of source water quality and flushing velocity on discharge water quality. The following parameters were evaluated: turbidity (Figure E.2), iron (Figure E.3), manganese (Figure E.4), and heterotrophic plate count (Figure E.5). Maximum values of these parameters from each valid bulk water flush were used to create the box-plots. The results indicate that the median values of each parameter are not significantly different (within a 95% confidence region around the median) between the two velocities. However, the median values of discharge iron and manganese were several orders of magnitude greater than the respective finished water concentrations. This difference represents the accumulation of iron and manganese deposits over time. However, based solely on these data, it was not possible to determine the source of the iron measured in the discharge water (i.e. finished water or from pipe material). It is likely that the manganese measured in the discharge water originated from the finished water, as cast and ductile iron pipe contain less than 0.5% manganese.

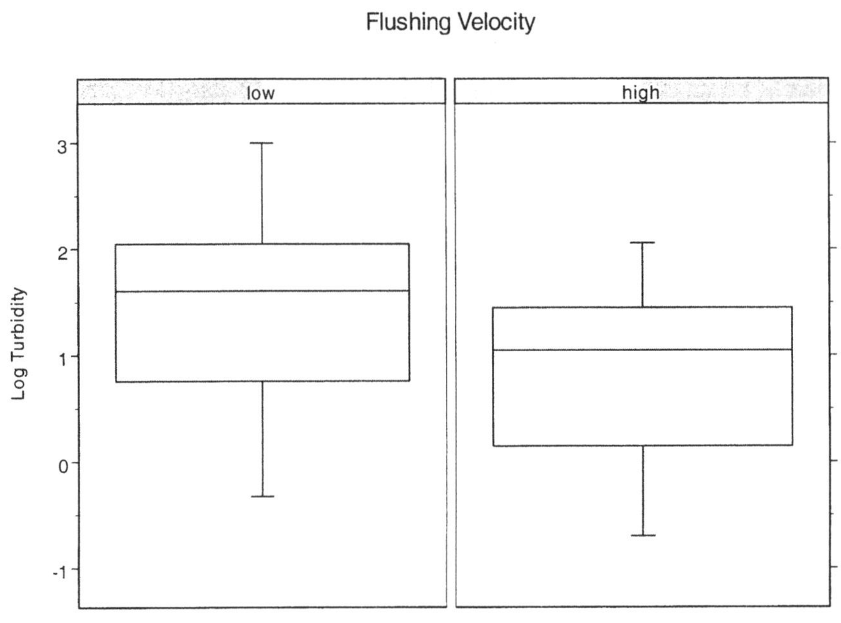

Figure E.2 Bulk water turbidity levels in low and high velocity discharge

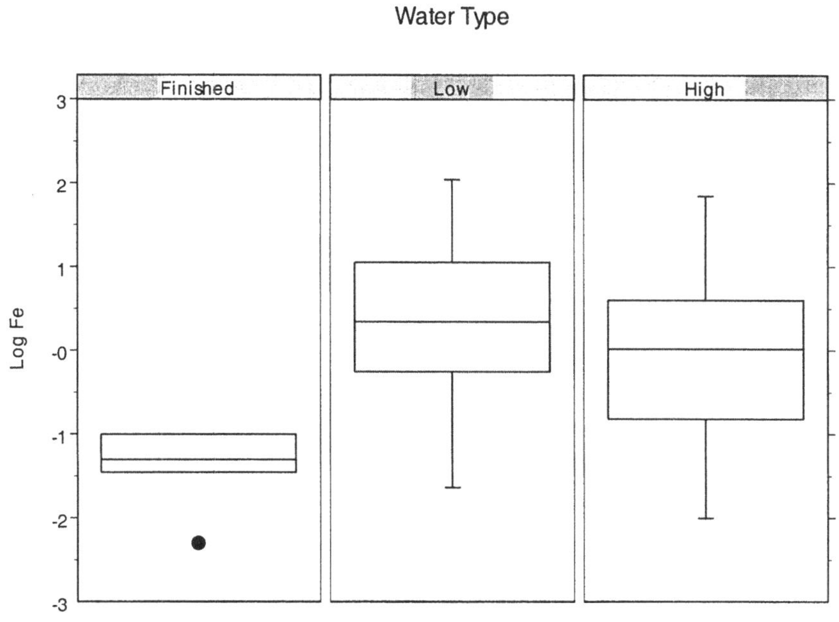

Figure E.3 Bulk water iron levels in finished water and low and high velocity discharge

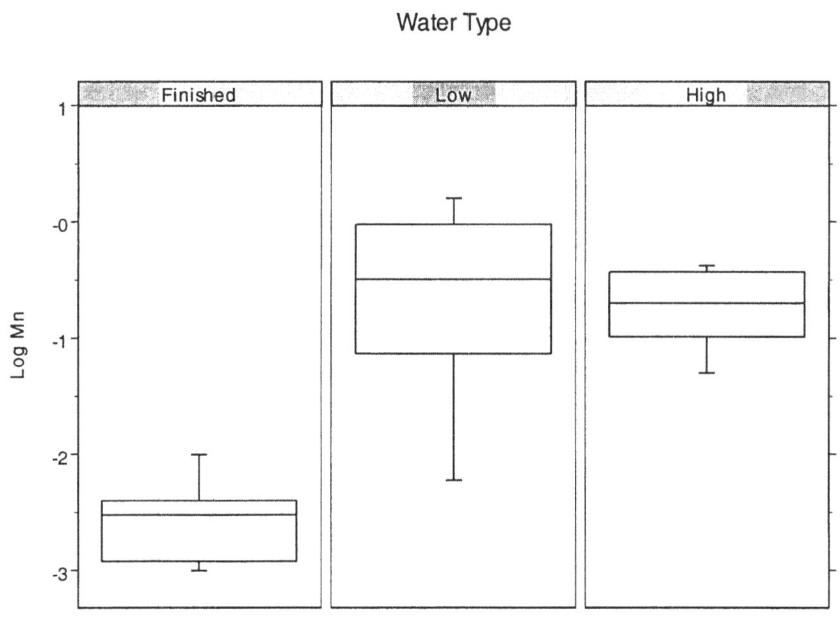

Figure E.4 Bulk water manganese levels in finished water and low and high velocity discharge

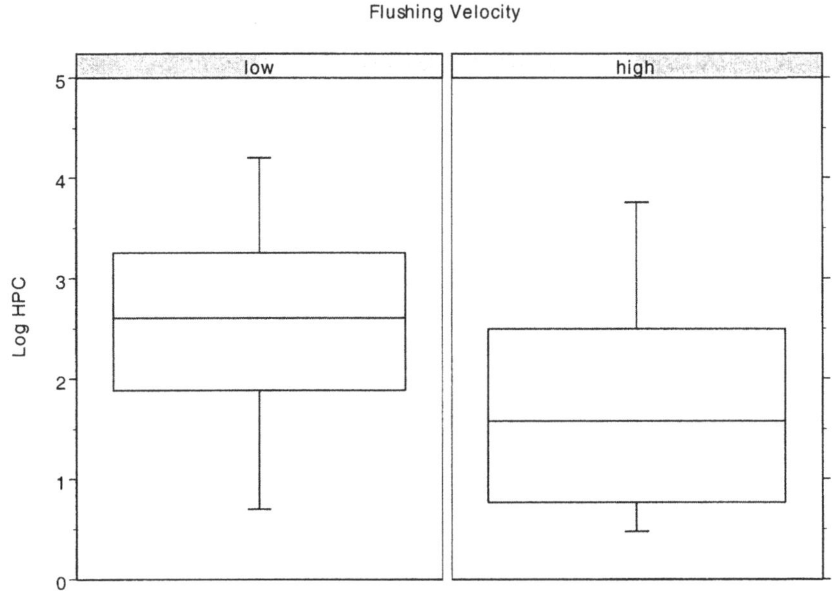

Figure E.5 Bulk water HPC levels in low and high velocity discharge

Box-plots were also generated to determine the impact of pipe material on discharge water quality. The following parameters were evaluated: turbidity (**Figure E.6**), iron (**Figure E.7**), and heterotrophic plate count (**Figure E.8**). Maximum values of these parameters from each valid bulk water flush were used to create the box-plots. For each of the bulk water parameters, the median values for DI, UCI, and AC pipe were not significantly different (based on an approximation of a 95% confidence region about the median). However, the median turbidity and iron levels collected from the field study with PVC piping were significantly less than those from the other pipes, particularly UCI. This result suggests that a significant source of iron particles may be unlined cast iron pipe (versus finished water contribution), as there was no UCI in the water system with PVC piping. Conversely, all field studies involving DI and AC pipe contained nearby UCI mains, which may have influenced (i.e., escalated) the iron levels in the discharge water. Median HPC levels in water collected from DI pipes were an order of magnitude higher than those observed in UCI pipes, although the rationale for this difference is unclear.

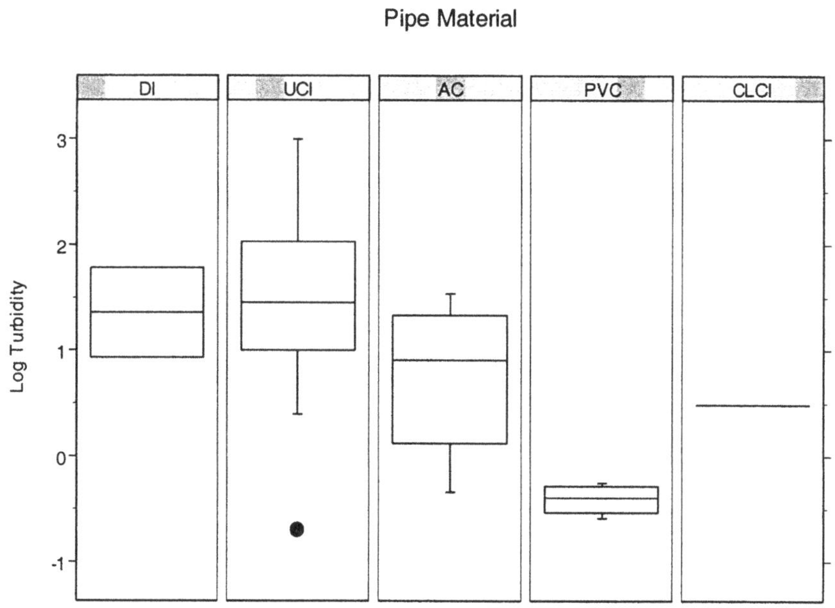

Figure E.6 Bulk water turbidity levels conditioned to pipe material

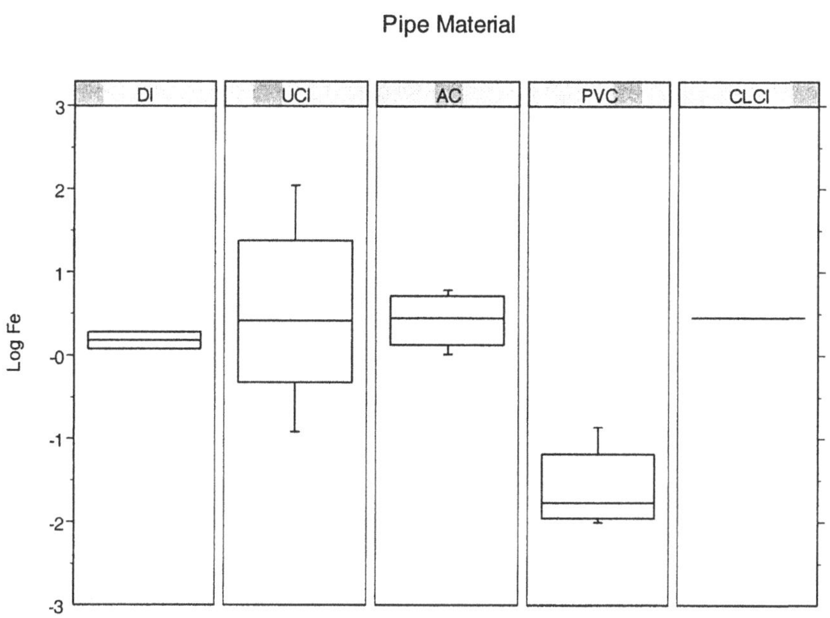

Figure E.7 Bulk water iron levels conditioned to pipe material

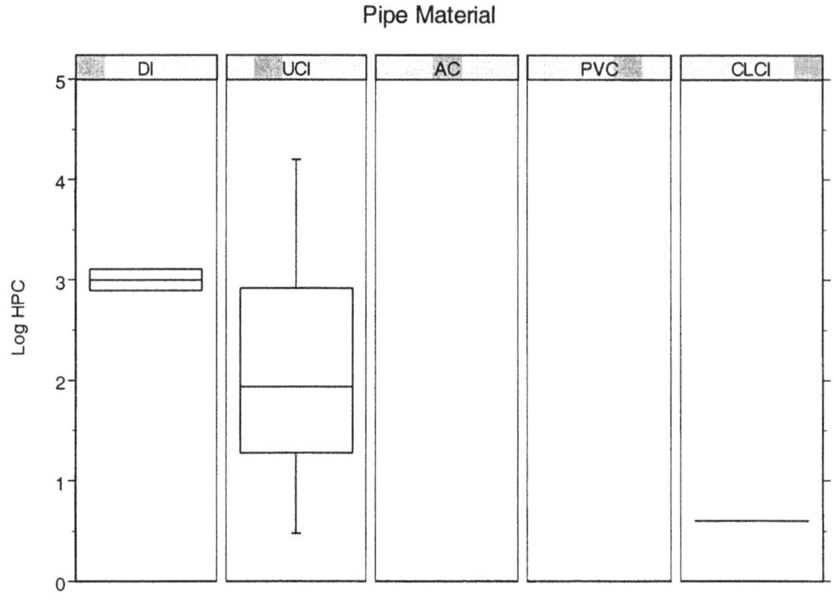

Figure E.8 Bulk water HPC levels conditioned to pipe material

REFERENCES

Abernathy, C.G., and A.K. Camper. 1998. The Effect of Phosphorus-Based Corrosion Inhibitors and Low Disinfectant Residuals on Distribution Biofilms. In *Proceedings of the AWWA Water Quality Technology Conference*. Denver, Colo.: AWWA.

Ackers, J., M. Brandt, and J. Powell. 2001. *Hydraulic Characterization of Deposits and Review of Sediment Modeling*. Report REF. No. 01/DW/03/18. London: UKWIR.

Ainsworth, R.G., J. Ridgway, and R.D. Gwilliam. 1979. Corrosion Products and Deposits in Iron Mains. *Pipes and Pipelines International*, 24(2):21-25 & 40.

Alasdair G. F. and R.M. Ashley. 1999. A Model for the Prediction and Control of Problematic Sediment Deposits. In *Proc. of the 8th International Conference on Urban Storm Drainage*. Reston, VA: ASCE.

Allen, M.J., R.H. Taylor, and E.E. Geldreich.1980. The Occurrence of Microorganisms in Water Main Encrustations. *Jour. AWWA*, 72:(11):614-625.

Anon. 1980. Practices for Identification of Crystalline Compounds in Water-Formed Deposits by X-ray Diffraction. D934-80. *Annual Book of ASTM Standards*, 11.02 (11). West Conshohoken, PA: ASTM.

Antoun, E.N., T. Tyson, and D. Hiltebrand. 1997. Unidirectional Flushing: A Remedy to Water Quality Problems such as Biologically Mediated Corrosion. In *Proceedings of the AWWA Annual Conference*. Denver, Colo.: AWWA.

AWWA and AWWARF. 1992. *Water Industry Database: Utility Profiles*. Denver, Colo.: AWWA.

AWWA and DVGW-Technologiezentrum Wasser. 1996. *Internal Corrosion of Water Distribution Systems*. Cooperative Research Report, 2nd Edition. Denver, Co.: AWWA.

Babaeyan-Koopaei K., D. Butler, and J. Davies. 1999. Modeling of Gross Solids Transport in Sewers. In *Proc. of the 8th International Conference Urban Storm Drainage*. Reston, VA: ASCE.

Barbeau, B., K. Julienne, V. Gauthier, R. Millette, and M. Provost. 1999. In *Proc. of the AWWA Water Quality Technology Conference*. Denver, CO: AWWA.

Baylis, J.R. 1926. Prevention of Corrosion and Red Water. *Jour. AWWA,* 15(6):598-631.

Benjamin, M.M., H. Sontheimer, and P. Leroy. 1996. Corrosion of Iron and Steel. In *Internal Corrosion of Water Distribution Systems, Co-operative Research Report.* 2nd Ed., Denver Colo.: AWWA and DVGW - Technologiezentrum Wasser.

Bhargava, D. and K. Rajagopal. 1992. An Integrated Expression for Settling Velocity of Particles in Water. *Wat. Res.*, 26(7):1005-1008.

Butler, G. & H.C.K. Ison. 1966. *Corrosion and its Prevention in Waters.* New York: Reinhold.

Camper, A.K., P. Butterfield, B. Ellis, and C. Abernathy. 1999. Coliforms in Biofilms: An Overview. In *Proceedings of the AWWA Water Quality Technology Conference*. Denver, Colo.: AWWA.

Cardew P., P. Marshall, J. Marshall, and W. Grayman. 2000. *Aspects of Modeling Chlorine in Water Distribution Systems*. London: UKWIR.

Carrière, A., B. Barbeau, V. Gauthier, C. Morissette, R. Millette, and A. Lalumière. 2002. Unidirectional Flushing: Loose Deposits Characterization in the Test Zones of Four Canadian Distribution Systems. In *Proceedings of the AWWA Water Quality Technology Conference*. Denver, Colo.: AWWA.

Chadderton, R.A., G.L. Christensen, and P. Henry-Unrath. 1992. *Implementation and Optimization of Distribution Flushing Programs*. Denver, Colo.: AWWA and AWWARF.

Clement, J., A. Sanvig, V. Snoeyink, W. Kriven, and P. Sarin.1998. Analyses and Interpretation of the Physical, Chemical, and Biological Characteristics of Distribution System Pipe Scales. In *Proceedings of the AWWA Water Quality Technology Conference*. Denver, Colo.: AWWA.

Crabtree, R.W. 1989. Sediments in Sewers. *Journal of the Institution of Water and Environmental Management*. WRc, Swindon.

DeRosa, S. 1993. *Loose Deposits in Water Mains*. Report No. DoE 3118-/2, WRc plc, Swindon.

DeRosa, S. and S. Miller. 1994. *Water Quality in Distribution*. Report No. UM 1430, WRc plc, Swindon.

Dombay, G., P. Piriou, and L. Kiene. 1999. Quantifying the Influence of Environmental Parameters on Bacterial Regrowth Phenomena in Distribution Systems. In *Proceedings of the AWWA Water Quality Technology Conference*. Denver, CO: AWWA.

Donlan, R.M., and W.O. Pipes. 1988. Selected Drinking Water Characteristics and Attached Microbial Population Density. *Jour. AWWA*, 80(11):70-76.

Durand R. and E. Condolios. 1952. The Hydraulic Transport of Coal and Solid Materials in Pipes. In *Proceedings of Colloquium on the Hydraulic Transport of Coal*. London: National Coal Board.

Economic and Engineering Services and Kennedy/Jenks/Chilton. 1989. *Economics of Internal Corrosion Control*. Denver, Colo.: AWWA and AWWARF.

Emde, K.M., K. Oberoi, and D. Smith. 1995. Evaluation of Various Methods for Distribution System Biofilm Control. In *Proceedings of the AWWA Annual Conference*. Denver, Colo.: AWWA.

Emde, K.M., K. Oberoi, and D. Smith.1997. Maximizing Distribution System Maintenance Activities. In *Proceedings of the AWWA Annual Conference*. Denver, Colo.: AWWA.

Emde, K.M., D.W. Smith, and R. Facey. 1992. Initial Investigation of Microbially Influenced Corrosion (MIC) in a Low Temperature Water Distribution System. *Water Research*, 26(2):169-175.

Evans, U.R. 1937. *Metallic Corrosion, Passivity and Protection*. London: Edward Arnold Ltd.

Folk, R.L. and W.C. Ward. 1957. Brazos River Bar, a Study in the Significance of Grain-Size Parameters. *J. Sed. Pet.*, 27: 3-27.

Francis, J.R.D. and P. Minton. 1984. *Civil Engineering Hydraulics*. 5th Ed. London: Edward Arnold Ltd.

Garde, R.J. and R.K.G. Ranga. 1985. *Mechanics of Sediment Transportation and Alluvial Stream Problems*. 2nd edition. Wiley Eastern Ltd.

Gauthier, V., B. Gerard, J.M. Portal, J.C. Block, and D. Gatel. 1999. Organic Matter as Loose Deposits in a Drinking Water Distribution System. *Water Research*, 33(4):1014-1026.

Gauthier, V., J.M. Portal, C. Rosin, J.C. Block, and D. Gatel. 1998. How Good are Water Distribution Systems for Transport of Particulate Matter? In *Proceedings of the AWWA Water Quality Technology Conference*. Denver, Colo.: AWWA.

Gauthier, V., C. Rosin, L. Mathieu, J.M. Portal, J.C. Block, P. Chaix, and D. Gatel. 1997. Characterization of the Loose Deposits in Drinking Water Distribution Systems. In *Proceedings of the AWWA Water Quality Technology Conference*. Denver, Colo.: AWWA.

Gilley, J.E., E.R. Kottwitz, G.A. Wieman. 1994. Hydraulic Conditions Required to Move Unanchored Residue Materials. *Jour. of Irrigation and Drainage Eng*, 120(3):591-606.

Gould, R.W. 1981. The Application of X-Ray Diffraction to the Identification of Corrosion Products. In *Proceedings of the AWWA Water Quality Technology Conference*. Denver, Colo.: AWWA.

Herson, D.S., D.R. Marshall, K.H. Baker, and H.T. Victoreen. 1991. Association of Microorganisms with Surfaces in Distribution Systems. *Jour. AWWA*, 83(7):103-106.

Herson, D.S., B. McGonigle, M.A. Payer, and K.H. Baker. 1987. Attachment as a Factor in the Protection of *Enterobacter cloacae* from Chlorination. *Applied and Environmental Microbiology*, 53(5):1178-1180.

Holt, D.M., T. Woodward, T. Rachwal, A. Delanoue, and J.S. Colbourne. 1994. A Pipe Based Experimental System to Study Factors Influencing Bacterial Growth in London's Water Mains. In *Proceedings of the AWWA Water Quality Technology Conference*. Denver, Colo.: AWWA.

Ihaka, R. and R. Gentelman, 1996. R: A Language for Data Analysis and Graphics. *J. Comp. Graph. Stat.*, 5(3): 299-314.

Jack, L. 1996. Prevention of Internal/External Corrosion - National Report UK. *Wat. Supply*, 14(3/4):286-287.

Joseph, S. and J.G. Pimblett. 2000. Unidirectional Flushing Program is Clean Sweep. *Opflow*, 26(1):1,6-7.

Kerneïs, A., S.C. Martin, and A. Deguin. 1996. Hydraulic Analysis for the Qualification of Distribution Networks. *Wat. Supply,* 14(3/4):515-521.

Kerr, C.J., K.S. Osborn, G.D. Robson, and P.S. Handley. 1998. Quantification of Biofilm Development on Distribution Pipe Materials. In *Proceedings of the AWWA Water Quality Technology Conference*. Denver, Colo.: AWWA.

Kirmeyer, G.J., W. Richards, and C. Smith. 1994. *An Assessment of Water Distribution Systems and Associated Research Needs*. Denver, Colo.: AWWA and AWWARF.

Klein H., and R. Forster. 1998. Network Operation Without Safety Chlorination. *Zurich Wat. Sup.*, 16(3/4):165-174.

Kölle, W. & H. Rösch. 1980. Examination of Encrustations of Drinking Water Pipes and Mineralogical Aspects. *Vom Wasser*, 55:159-177.

Larsen, T. 2000. Measuring Variation of the Apparent Settling Velocity for Fine Particles. *Wat. Res.*, 34(4):1417-1418.

Latimer, P. 1996. Network Management Systems - National Report UK. *Wat. Supply*, 14(3/4):121-155.

LeChevallier, M.W., T.M. Babcock, and R.G. Lee. 1987. Examination and Characterization of Distribution System Biofilms. *Applied and Environmental Microbiology*, 53(12):2714-2724.

LeChevallier, M.W., N.J. Welch, and D.B. Smith. 1996. Full-Scale Studies of Factors Related to Coliform Regrowth in Drinking Water. *Applied and Environmental Microbiology*, 62(7):2201-2211.

Macke, E. 1982. About Sedimentation at Low Concentrations in Partly Filled Pipes. Mitteilungen, Leichtweiss - Institut Fur Wasserbau de Technischen Universitat Braunschweig, Heft 76, pp 1-151.

Maier, S. 1998. Modelling Water Quality for Water Distribution Systems PhD thesis. Brunel University.

Marshall, G.P. 2000. *Understanding and Preventing Discolored Water.* UKWIR Report No. 01/DW/03/17. London: UKWIR Ltd.

May, R.W.P. 1994. Appendix 2, Section A2.2.9, in *Design of Sewers to Control Sediment Problems – Report 141.* London: CIRIA.

McMath, S. M., A. Delanoue, D.M. Holt, S. Maier, and C.A. Woodward. 1997. 'Clumps' Shed From Pipe Walls in Water Distribution Systems. *In Proceedings of AWWA Water Quality Technology Conference.* Denver, Colo.: AWWA.

Nagy, L.A., and B.H. Olson. 1985. Occurrence and Significance of Bacteria, Fungi and Yeasts Associated With Distribution Pipe Surfaces. In *Proceedings of the AWWA Water Quality Technology Conference.* Denver, Colo.: AWWA.

Nalluri, C. and E.M. Alvarez. 1992. The Influence of Cohesion on Sediment Behavior. *Wat. Sci. Tech.*, 25(8).

Nalluri, C., and W. Dabrowski. 1994. Need for New Standards to Prevent Deposition I Wastewater Sewers. *Jour. of Environmental Engineering*, 120(5):1032-1043.

Oberoi, K. 1994. Distribution Flushing Program: The Benefits and Results. In *Proceedings of the AWWA Annual Conference.* Denver, Colo.: AWWA.

Pattison, P.L. 1980. Conducting a Regular Main Flushing Program. *JAWWA*, 72(2): 88-90.

Peyton, B.M. 1996. Effects of Shear Stress and Substrate Loading Rate on *Pseudomonas Aeruginosa* Biofilm Thickness and Density. *Water Research*, 30(1):29-36.

Prince, R., K. McManus, and I. Goulter. 2000. Color, Turbidity Levels and Dirty Water Customer Complaints - Water System Performance Indicators? In *Proc. of The World Water Congress.*

Rasmussen, M.R. 1997. Solid Dynamics in Secondary Settling Tanks. PhD thesis. Aalborg University.

Ridgway, H.F., E.G. Means and B.H. Olson. 1981. Iron Bacteria in Drinking-Water Distribution Systems: Elemental Analysis of *Gallionella* Stalks, Using X-Ray Energy-Dispersive Microanalysis. *Applied and Environmental Microbiology*, 41(1):288-297.

Ridgway, H.F. and B.H. Olson, B.H. 1981. Scanning Electron Microscope Evidence for Bacterial Colonization of a Drinking-Water Distribution System. *Appl. Environmental Microbiology*, 41(1):274-287.

Rijkens, B.A. 1967. Verhalten der polyphosphate bei der korrosion der metalle. *Wasser, Luft, Betr.*, 11:728-733.

Saffman, P.G. 1965. The Lift on a Small Sphere in a Slow Shear Flow. *J. Fluid Mechanics*, 22:385-400.

Sakakibara, T. 1996. Sediments Flushing Experiment in a Trunk Sewer *Wat. Sci. Tech.*, 33(9):229-235.

Schock, M.R. 1990. Internal Corrosion and Deposition Control - Chapter 17. In *Water Quality and Treatment, A Handbook of Community Water Supplies.* Denver, Colo.: AWWA.

Sears, F.W., M.W. Zemansky, and H.D. Young. 1988. *University Physics.* Fifth Edition. Boston: Addison-Wesley.

Sethi, V., P. Biswas, R. Clark, and L. Rossman. 1999. Particle Deposition and Re-Entrainment in Drinking Water Distribution Systems. In *Proceedings of the AWWA Water Quality Technology Conference.* Denver, Colo.: AWWA.

Skipworth, P., S. Tait, and A. Saul. 1996. Laboratory Investigations Into Cohesive Sediment Transport in Pipes. *Wat. Sci. Tech.*, 33(9):187-193.

Slaats, N., ed. 2001. Processes Involved in the Generation of Discolored Water. Nieuwegein, The Netherlands: KIWA.

Smith. S.E., J.S. Colbourne, D.M. Holt, B.J. Lloyd, and A. Bisset. 1997. An Examination of the Nature and Occurrence of Deposits in a Distribution System and Their Effect on Water Quality. In *Proceedings of the AWWA Water Quality Technology Conference*. Denver, Colo.: AWWA.

Smith, C.D., J.F. Smith, and B. Milosky. 1996. Excessive Loss of Chloramine Residual: A Case Study. In *Proceedings of the AWWA Annual Conference*. Denver, Colo.: AWWA.

Smith, S.E., T. Ta, D.M. Holt, A. Delanoue, and J.S. Colbourne. 1998. Minimizing Red Water in Drinking Water Distribution Systems. In *Proceedings of the AWWA Water Quality Technology Conference*. Denver, Colo.: AWWA.

Smith, S. E., T. Ta, D.M. Holt, A. Delanoue, J.S. Colbourne, H.L. Chamberlain, and B.J. Lloyd. 1999. A Pipeline Testing Facility for the Examination of Pipe-Wall Deposits and Red-Water Events in Drinking Water. *Water and Environmental Management*, 13(1).

Sontheimer, H., W. Kölle, and V.L. Snoeyink. 1981. The Siderite Model of the Formation of Corrosion-Resistant Scales. *Jour. AWWA*, 73(11):572-579.

Stephenson, G. 1989. *Removing Loose Deposits From Water Mains: Operational Guidelines - A Source Document for the Water Mains Rehabilitation Manual*. WRc plc, Swindon.

Talbot, L., R.K. Cheng, R.W. Schefer, and D.R. Willis. 1980. Thermophoresis of Particles in a Heated Boundary Layer. *J. Fluid Mechanics*, 101: 737-758.

Tuovinen, O.H. & E. Nurmiaho. 1979. Microscopic Examination of Bacteria in Fe (III)-Oxide Deposited from Ground Water. *Microbial Ecology*, 5:57-66.

Tuovinen, O.H., K.S. Button, A. Vuorinen, L. Carlson, M. Mair, L.A. Yut. 1980. Bacterial, Chemical and Mineralogical Characteristics of Tubercles in Distribution Pipelines. *Jour. AWWA*, 72(11): 626-635.

Valentine, R.L., B. Angerman, S. Hackett, P. Vikesland, and S. Slattenow. 1998. Characterization of Disinfectant Decay and DBP Formation in the Presence of Water Distribution System Deposits. In *Proceedings of the AWWA Water Quality Technology Conference*. Denver, Colo.: AWWA.

Van den Hoven, T., and J.H.G. Vreeburg. 1992. Distribution System Analysis by Continuous Monitoring and Network Calculation *Wat. Supply*, 10:117-124.

Van den Hoven, T., D. van der Kooij, J. Vreeburg, and H. Brink. 1994. Methods to Analyze and to Cure Water Quality Problems in Distribution Systems. *Wat. Supply*, 12(3/4):151-159.

van der Kooij, D., J.H.M. van Lieverloo. 1998. *Distributing Drinking Water With Low or Zero Disinfectant Residual*. Nieuwegein, The Netherlands: KIWA.

van Lieverloo, J.H.M., R. Buuren, G. Veenendaal, and D. van der Kooij. 1998. Controlling Invertebrates in Distribution Systems With Zero or Low Disinfectant Residual. *Wat. Supply*, 16(3/4):199-204.

van Lieverloo, J.H.M., D. van der Kooij, and G. Veenendaal. 1994. National Survey of Invertebrates in Drinking Water Distribution Systems in the Netherlands. In *Proceedings of the AWWA Water Quality Technology Conference*. Denver, Colo.: AWWA.

Vibert, A., P. Gayaud, and D. Amyot. 1991. Une unite mobile de diagnostic et d'evaluation des depots dans le reseaux de distribution *L'eau, L'industrie, Les Nuisances*, 149:49-52.

Vreeburg, J.H.G.: 1996. Red water, cleaning, cause and consequences. Kiwa report SEW 96.008, Nieuwegein, The Netherlands.

Walker, A.P. 1983. The Microscopy of Consumer Complaints. *J. Inst. Wat. & Envir. Mgt.*, 37(3): 200-214.

Walski, T.M. 1991. Understanding Solids Transport in Water Distribution Systems. In *Water Quality Modeling in Distribution Systems*. Cincinnati, OH. AWWARF.

Walski, T.M. and S.J. Draus. 1996. Predicting Water Quality Changes During Flushing. In *Proceedings of the AWWA Annual Conference*. Denver, Colo.: AWWA.

Weidmann, G. 1970. Aufbau und chemische zusammensetzung von rosten in eisernen rohrleitungen. *Gas-Wasserfach, Wasser-Abwasser*, 111:713-715.

Wentworth, K.C. 1922. A Scale Grade and Class Terms for Classic Sediments. *J. Geol.*, 30:377-392.

Williams, S.M. 2002. Mains Flushing and Sediment Measurement in Water Mains. In *Proceedings of the AWWA Water Quality Technology Conference*. Denver, Colo.: AWWA.

Williams, S.M. 2000. *Practical Guidelines for Mains Flushing*. Mayfield Consulting Ltd.

Woodward, C.A., T. Ta, J.S. Colbourne, and D.M. Holt. 1995. Behavior of Particles in a Large-Scale Experimental Pipe System. In *Proceedings of the AWWA Water Quality Technology Conference*. Denver, Colo.: AWWA.

Xu, Y.L. 1996. Gross Solids Movement in Sewers. PhD thesis. University of Westminster.

Zug, M., L. Bellefleur, L. Phan, and O. Scrivener. 1998. Sediment Transport Model in Sewer Networks - a New Utilization of the Velikanov Model. *Wat. Sci. Tech.*, 37(1):187-196.

ABBREVIATIONS

AC	asbestos cement
AK	Alaska
Al	aluminum
ANC	Anchorage, Alaska
AWWA	American Water Works Association
AwwaRF	Awwa Research Foundation
BEL	Bellevue Utilities Department, Bellevue, Washington
C	Celsius
Ca	Calcium
CA	California
$CaCO_3$	calcium carbonate
CFD	computational fluid dynamics
cfu	colony forming unit
CFU/cm2	colony forming unit per centimeter squared
CFU/mL	colony forming unit per milliliter
CIN	Cincinnati, Ohio
CLCI	cement-lined cast iron
CLDI	cement-lined ductile iron
CLE	Cleveland, Ohio
Colo	Colorado
DI	ductile iron
DOV	Dover, New Hampshire
E_i	Low velocity mass removal efficiency for parameter i (%)
F	Fahrenheit
Fe	iron
Fps	feet per second
Ft	foot
ft/min	feet per minute
G	gram
GAC	granular activated carbon
Gpm	gallons per minute
HPC	heterotrophic plate count
in	inches
IR	Irvine Ranch Water District, Irvine, California

Km	kilometer
L/s	liters per second
m	meter
M_i^*	Sum of mass removed at low and high velocities (mg)
MCL	Maximum Contaminant Level
MDPE	medium density polyethylene
ME	Maine
mg	milligrams
mg/ft^2	milligrams per square feet
mg/kg	milligrams per kilogram
mg/L	milligrams per liter
m^3/hr	cubic meters per hour
m/hr	meters per hour
min	minute
mL	milliliter
mm	millimeter
Mn	manganese
M/s	meters per second
NA	Not available
N/a	not applicable
ND	Not detected
NH	New Hampshire
N/m^2	Newtons per meter squared
NNH	Newport News, Virginia's Hilton Village system
NNI	Newport News, Virginia's Ivy Farms system
No/mL	number per milliliter
N.S.	not specified by author
NT	not tested
NTU	nephalometric turbidity units
O&M	operation and maintenance
OH	Ohio
OR	Oregon
P	phosphorous
PAC	Project Advisory Committee
Pb	lead
POR	Portland, Oregon
Proc.	Proceedings
PVC	polyvinyl chloride
S	Sulfur
SB	South Berwick, Maine

Si	Silicon
SCAL	Southern California Water Company
SEM	scanning electron microscopy
SG	specific gravity
SW	southwest
TH	Thames Water
TSS	total suspended solids
UCI	unlined cast iron
UDF	unidirectional flushing
μg	microgram
μg/L	micrograms per liter
UK	United Kingdom
UKWIR	United Kingdom Water Industry Research
µm	micrometer
UPVC	unplasticized polyvinyl chloride pipe
U.S.	United States
USEPA	United States Environmental Protection Agency
VA	Virginia
VOC	volatile organic compound
WA	Washington
WPCF	Water Pollution and Control Federation
XRD	x-ray diffraction
yr	year
YVR	Yarra Valley R
YVW	Yarra Valley Water, Melbourne, Australia

IWA Publishing's authorised EU representative for General Product Safety Regulations is Diane D'Arras, 15 rue Duret, 75116 Paris, France, e-mail: safety@iwap.co.uk.

Printed and bound by CPI Group (UK) Ltd, Croydon, CR0 4YY
17/09/2026
02218810-0003